Protocols and Applications in Enzymology

Protocols and Applications in Enzymology

Seema Anil Belorkar

Assistant Professor,
Microbiology and Bioinformatics Department,
Atal Bihari Vajpayee University,
Bilaspur (C.G), India

Sudisha Jogaiah

Assistant Professor,
Laboratory of Plant Healthcare and Diagnostics,
PG Department of Studies in Biotechnology and Microbiology,
Karnatak University, Pavate Nagar, Dharwad, Karnataka, India

Academic Press is an imprint of Elsevier
125 London Wall, London EC2Y 5AS, United Kingdom
525 B Street, Suite 1650, San Diego, CA 92101, United States
50 Hampshire Street, 5th Floor, Cambridge, MA 02139, United States
The Boulevard, Langford Lane, Kidlington, Oxford OX5 1GB, United Kingdom

Library of Congress Cataloging-in-Publication Data
A catalog record for this book is available from the Library of Congress

British Library Cataloguing-in-Publication Data
A catalogue record for this book is available from the British Library

ISBN: 978-0-323-91268-6

For information on all Academic Press publications visit our website at https://www.elsevier.com/books-and-journals

Publisher: Andre Gerhard Wolff
Acquisitions Editor: Michelle Fisher
Editorial Project Manager: Sam W. Young
Production Project Manager: Niranjan Bhaskaran
Cover Designer: Alan Studholme

Typeset by TNQ Technologies

Contents

Preface

Protocols and applications in enzymology is a book presenting intensive, coordinated, and synchronized information with the perfect blend of historical information, thrust areas in research, and the resultant protocols much desired in the academics and research labs.

This book is a one-stop solution for students, academicians, research beginners, and entrepreneurs. The initial two chapters reveal the glorious history of enzymology with the pioneering research and its deep-rooted connection with the discoveries by eminent chemists. It explains the diversity an enzyme molecule exhibits and their implications on industrial processes.

The book describes various stages of screening technologies, conventional Vs modern, fermentations, and their types and provides exclusive research-based protocols for industrially important enzymes which are desperately required in research labs and industrial research units. Thus, this book is a link to protocols desired by the industries as the intensive outcome of the research community. It also elucidates the "Waste to value" term by discussing conversion of the trapped energy in wastes into bioactive molecules.

The book will hopefully update the readers from agricultural to industrial sector with recent advances in usage of enzymes as frontier tools and finally focusing on applications of enzymes in different walks of life and future applications.

(Dr. Seema Anil Belorkar and Dr. Sudisha Jogaiah)
Authors

CHAPTER

Enzymes—past, present, and future

1

1.1 General introduction

Enzymes today need no introduction. They are well recognized and acknowledged in every walk of life from a layman to a research person entangled in molecular biology protocols. These magical molecules have proved to be an efficient tool in almost all industries directly or indirectly.

The term "Enzyme" was proposed by Wilhelm Kuhne, which in Greek means "leavened" or "in yeast" in 1877. Enzymes have gained recognition as a Biological catalyst, which is generally proteins with the exceptions of ribozymes. The nature and activity of any enzyme depend upon the sequence of amino acids of the polymer. The enzymes can be active as a single chain (monomeric protein) or can have a number of polymeric chains in form of a functional complex (oligomeric protein). The function of an enzyme is dependent on its structure and conformation that in turn depends on the amino acid sequence (Berg et al., 2002). The structure of any enzyme is a very sensitive and adaptive feature in context to its changing environment. It cannot be the sole feature governing its function (Somero, 1978).

The basic development of enzymology is accredited to pioneering observations in the field that are enlisted in Table 1.1.

1.2 Characteristics of enzymes

Enzymes are biological catalysts. They catalyze biological reactions in a similar manner as Chemical catalysts do. The enzymes transform their substrates (reactants) to products under normal conditions of temperature and pressure in contrast to chemical catalyst. Fig. 1.1 elucidates the general path of enzyme action. It is a two-step reaction. The first step is reversible and forms an unstable ES complex. If the substrate is specific, the reaction progresses for the second step; otherwise, the breakdown causes the release of the substrate and a backward reaction takes place. The second step is irreversible and progresses only in the forward direction when the enzyme finds its specific substrates.

Protocols and Applications in Enzymology. https://doi.org/10.1016/B978-0-323-91268-6.00007-7

Table 1.1 Pioneering contributions in enzymology.

Year	Noble laureates	Contributions
2018	Frances H. Arnold	The directed evolution of enzymes
	George P. Smith Sir Gregory P. Winter	The phage display of peptides and antibodies
2017	Jacques Dubochet Joachim Frank Richard Henderson	Developing cryo-electron microscopy for the high-resolution structure determination of biomolecules in solution
2016	Jean-Pierre Sauvage Sir J. Fraser Stoddart Bernard L. Feringa	The design and synthesis of molecular machines
2015	Tomas Lindahl Paul Modrich Aziz Sancar	Mechanistic studies of DNA repair
2012	Robert J. Lefkowitz Brian K. Kobilka	Studies of G-protein-coupled receptors
2009	Venkatraman Ramakrishnan Thomas A. Steitz Ada E. Yonath	Studies of the structure and function of the ribosome
2008	Osamu Shimomura Martin Chalfie Roger Y. Tsien	The discovery and development of the green fluorescent protein, GFP
2006	Roger D. Kornberg	The molecular basis of eukaryotic transcription
2002	John B. Fenn Koichi Tanaka	Development of soft desorption ionization methods for mass spectrometric analyses of biological macromolecules
	Kurt Wüthrich	Development of nuclear magnetic resonance spectroscopy for determining the three-dimensional structure of biological macromolecules in solution
1997	Paul D. Boyer John E. Walker	Elucidation of the enzymatic mechanism underlying the synthesis of adenosine triphosphate (ATP)
	Jens C. Skou	For the first discovery of an ion-transporting enzyme, Na+, K+-ATPase
1993	Kary B. Mullis	Invention of the polymerase chain reaction (PCR) method
	Michael Smith	Oligonucleotide-based, site-directed mutagenesis and its development for protein studies
1990	Elias James Corey	The theory and methodology of organic synthesis
1989	Sidney Altman Thomas R. Cech	Discovery of catalytic properties of RNA
1975	John Warcup Cornforth	The stereochemistry of enzyme-catalyzed reactions
	Vladimir Prelog	The stereochemistry of organic molecules and reactions
1974	Paul J. Flory	The physical chemistry of the macromolecules

Table 1.1 Pioneering contributions in enzymology.—*cont'd*

Year	Noble laureates	Contributions
1972	Christian B. Anfinsen Stanford Moore William H. Stein	Ribonuclease, especially concerning the connection between the amino acid sequence and the biologically active conformation The connection between chemical structure and catalytic activity of the active center of the ribonuclease molecule
1970	Luis F. Leloir	Discovery of sugar nucleotides and their role in the biosynthesis of carbohydrates
1969	Derek H. R. Barton Odd Hassel	The development of the concept of conformation and its application in chemistry
1965	Robert Burns Woodward	Achievements in the art of organic synthesis
1964	Dorothy Crowfoot Hodgkin	X-ray techniques of the structures of important biochemical substances
1963	Karl Ziegler Giulio Natta	The chemistry and technology of high polymers
1962	Max Ferdinand Perutz John Cowdery Kendrew	The structures of globular proteins
1961	Melvin Calvin	The carbon dioxide assimilation in plants
1958	Frederick Sanger	Structure of proteins, especially that of insulin
1957	Lord (Alexander R.) Todd	Nucleotides and nucleotide coenzymes
1956	Sir Cyril Norman Hinshelwood Nikolay Nikolaevich Semenov	Mechanism of chemical reactions
1955	Vincent du Vigneaud	The first synthesis of a polypeptide hormone
1953	Hermann Staudinger	Macromolecular chemistry
1950	Otto Paul Hermann Diels Kurt Alder	Discovery and development of the diene synthesis
1946	James Batcheller Sumner John Howard Northrop Wendell Meredith Stanley	Enzymes can be crystallized Preparation of enzymes and virus proteins in a pure form
1929	Arthur Harden Hans Karl August Simon von Euler-Chelpin	The fermentation of sugar and fermentative enzymes
1926	The (Theodor) Svedberg	Disperse systems
1909	Wilhelm Ostwald	Catalysis and for his investigations into the fundamental principles governing chemical equilibria and rates of reaction
1907	Eduard Buchner	Biochemical researches and his discovery of cell-free fermentation
1902	Hermann Emil Fischer	Recognition of the extraordinary services he has rendered by his work on sugar and purine syntheses

FIGURE 1.1

General mechanism of enzyme-catalyzed reaction.

1.2.1 Enzyme as a molecule

Enzymes are macromolecules having a high molecular weight (MW) and are disproportionately bigger when compared to their substrate. Generally, the MW of enzymes covers a broad range. The general properties in relation to its functions are explained in Fig. 1.2.

Dixon and Web (1958) tabulated enzymes in three series or classes ($2n \times 12{,}000$, $2n \times 16{,}000$, and $2n \times 19{,}000$), where n is integral 0–4. A hypothesis was proposed by Wright (1962) using experimental data but as the groups of enzymes were formed, there remained a question of Biasedness. Johnston et al. (1945) proposed a better approach of the correlation function for validation of the Svedberg hypothesis (Svedberg and Pedersen, 1940). The MW of proteins is calculated by Svedberg's equation depending on the force applied and the mass of the sedimenting molecule.

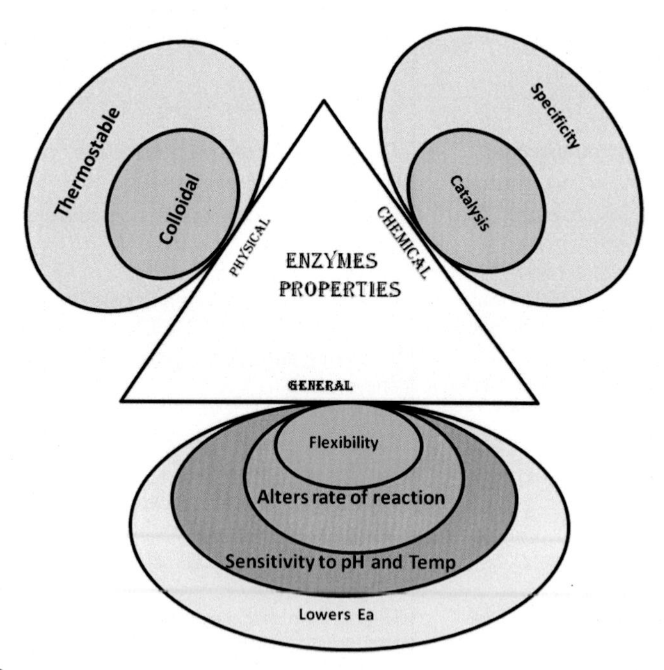

FIGURE 1.2

Properties and functions of enzymes.

The sedimentation velocity was fundamentally used for the determination of molecular weights of proteins. Presently, an approach using Gel filtration Chromatography and SDS-PAGE are the advanced methods for the MW determination.

As we all know that most enzymes are proteins, their general behavior is colloidal as an attribute of their high molecular weight. Enzymes that are proteinaceous are thermolabile in nature. Generally, enzymes are active at room temperature, at low temperatures they are reversibly inhibited, and at higher temperatures they are irreversibly inactivated.

Although, this temperature sensitivity can be handled, if enzymes are stored in the form of dried extracts. The stability of the enzymes increases along with increased shelf life. Every enzyme requires amicable pH and temperature conditions to explicit its highest activity referred as optimum pH and temperature. In zones before and after the optimum zone, the enzyme activity decreases. Fig. 1.3 explains the maximum enzyme activity in response to optimum pH and temperature conditions.

1.2.2 Catalytical property

A most striking feature of the enzyme is its catalytic power that makes it a robust synthetic tool. In comparison to chemical catalysts, enzymes are wonderful because they catalyze the transformation of substrates at normal temperature and pressure available inside the biological system. Enzyme caters high potential of substrate conversion as compared to reactions that run by themselves. It remains unaltered after the reaction is complete, providing the alternative of its recycling for better yield.

1.2.3 Specificity

The uniqueness of the enzyme is its structural conformation that renders the quality of being an effective catalyst. The catalysis conferred by the enzyme is very focused

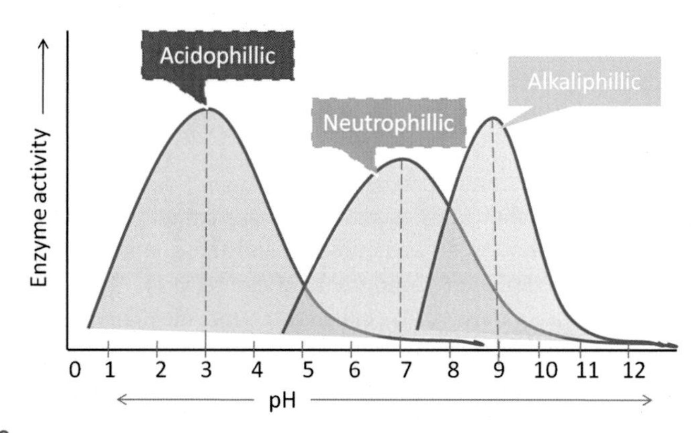

FIGURE 1.3

Sensitivity of the enzymes toward pH and temperature.

Table 1.2 Different types of specificities exhibited by enzymes.

Types of enzyme specificities		
Bond specificity (relative specificity)	**Enzyme** Peptidase Lipase	**Action on** Peptide bond Ester bond
Group specificity (structural specificity)	**Enzyme** Pepsin	**Action on** Peptide bond linked to aromatic amino acids like phenylalanine, tyrosine, and tryptophan
Substrate specificity (absolute specificity)	**Enzyme** Sucrase Arginase Carbonic anhydrase o	**Action on** Sucrose Arginine carbonic acid.
Optical specificity (stereo-specificity)	**Enzyme** L-amino acid oxidase α-amylase	**Action on** L-amino acids α-1-4 glycosidic linkage of starch and glycogen
(Cofactor specificity)	**Enzyme** Glucose-6-phosphate dehydrogenase	**Cofactor** NADH
(Geometric specificity)	**Enzyme** Alcohol dehydrogenase can oxidize Chymotrypsin	**Substrates** Methanol and *n*-propanol to aldehydes. Peptide bond and ester bond

and confined to either one or a group of structurally related molecules. The specificity can be classified into absolute specificity and relative specificity. Nowadays, enzymes are finding immense applications on grounds of Regioselective and Stereoselective (Mu et al., 2020). Table 1.2 explains the specificities of the different enzymes along with examples falling in each category.

1.2.4 Holoenzyme

Many enzymes have a nonprotein part along with them. Such conjugated enzymes found associated with another kind of molecules are called the holoenzyme. The protein component is called as the apoenzyme, and the nonprotein component is called the cofactor. If the cofactor is inorganic in nature, it is referred as prosthetic group and if it is organic in nature, it is called as coenzyme. The prosthetic groups are intensely associated with the apoenzyme and are not easily available. Contrarily, the coenzymes are very loosely bound resulting in easy separation from the protein component.

The cofactor is a helper micromolecule that confers enzyme, its functionality. The nonprotein nature makes it thermostable. The main role of cofactor is it becomes a significant part in the reaction center of the enzyme for catalysis involving the removal of functional groups. Cofactors have the ability to get associated with other enzymes and help their functionality after association.

1.2.5 Turnover number

As we all know, enzymes are biological catalysts that remain unaltered after the reaction is complete. The rate of enzyme catalyst reactions depends upon the number of substrate molecules converted into the product in a given specified time. Turnover number is therefore defined as the amount of substrate converted into the product in 1 s, in the given specified condition (where the concentration of substrate is in saturation). Enzymes exhibit a wide variety of turnover numbers ranging from 0.5 to 10^5.

1.2.6 Reversibility

When we focus on catalyzed reactions, the enzyme catalyzed reactions differ from the chemical catalyst in a way that it functions according to the need of the biological system. The nature of the enzyme varies in accordance with the role it is executing in the metabolism. Therefore, enzyme-catalyzed reactions include unidirectional, bidirectional types, and cvcn ccgulatory reactions.

1.2.7 Weight of enzyme

Most of the enzymes range between 20 kD and 60,000 kD molecular weight. The 3-D structure is visualized in terms of a macromolecule with 327 nm in size and eventually appears to be very large in context to the cellular environment. The review of the enzyme revealed the large size of the enzyme offers a large surface area of the molecule to circumvent the binding sites of substrates with appropriate orientation.

There are reports of two esterases very small molecular weight enzymes from *Candida lipolytica* and *Bacillus stearothermophilus* with 5.7 kDa (56 amino acid residue) and the *Bacillus* enzyme is 1.57 kDa (17 amino acids) (Mattey et al., 1998). The highest MW enzymes reported are from *Y. lipolytica* with a molecular weight of 120 kDa, the 90 kDa leucine aminopeptidase II from *Aspergillus oryzae* (Nicaud et al., 2002), and glucoamylase from *Arxula adeninivorans* (Swennen et al., 2002).

1.2.8 Sensitivity

Yet another feature of enzymes is that they are very sensitive to the environment in which they are present. The environmental conditions impose tremendous influence on the three-dimensional structure of the enzyme molecule and hence affect its activity and catalytic power to the extent whether it will remain in active state or be inactive.

The pH, temperature, and concentration of ions are few factors that impede the biological functioning of the molecule. These consequences are due to the disturbance of some vital bonds required for the architectural makeup of the catalytic site or the active site of the enzyme, rendering it inactive or denatured. Every enzyme has a unique sense of tolerance toward a particular limit of pH and temperature or concentration of ions. The tolerance capacity of enzymes is naturally inherited by the enzyme from the environment it belongs.

1.2.9 Active site

Most of the enzymes are high molecular weight molecules, and the majority of their monomeric amino acids are for maintaining their three-dimensional conformation. They play a structural role. The functional role is played by few amino acids which participate in the catalytically active center.

The three-dimensional conformation creates events or crevices that are narrow and stereospecific for the substrate. The dimensional congruence of the active site and the substrate structure is the foundation for the wonderful specificity phenomena observed between the enzyme and the substrate. Although the substrate-binding site and the catalytic centers are different, they reside close to one another inside the microcatalytic environment of the enzyme molecule. The regulatory enzymes possess an additional site for binding of the modulator molecule to execute positive or negative regulatory response as the case may be.

1.3 The concept of cell-free fermentation

Enzyme received acknowledgment due to Edward Buchner (1897) with the Founder enzyme for which he received Nobel Prize. He was the first person to use the prefix of the substrate to name the enzyme—for example, protease acts upon proteins as substrate, lipase on lipids as substrate, and so forth. Alternative nomenclature depends on the type of reaction catalyzed such as polymerase, dehydrogenase, and oxidoreductase. Thus, the initial pattern of nomenclature depended upon either the type of the substrate or the type of the reaction used as a prefix for ase.

1.4 Nomenclature of enzymes

A system for the nomenclature of enzymes was developed by the International Union of Biochemistry and Molecular Biology. The enzymes are assigned a Commission number called as the EC number. The enzymes are divided into six classes (Florkin and Stotz, 1965), but in 2018, Class 7 was introduced for Translocases (Tipton, 2018). Every enzyme is designated a four number sequence, which is called as the EC number for the enzyme. In the Commission number, every digit signifies information regarding the enzyme classification. The current classes of enzymes are categorized in Table 1.3.

Table 1.3 Enzyme nomenclature with recent advances.

Class of enzyme	Name of class	Reaction catalyzed
EC 1	Oxidoreductases	Oxidation/reduction reaction
EC 2	Transferases	Transfer or exchange of certain groups between substrates
EC 3	Hydrolases	Substrate hydrolysis
EC 4	Lyases	Removal of a group or its reverse reaction
EC 5	Isomerases	Interconversion of isomers
EC 6	Ligases	Synthesis of one product by combining two molecules of substrate (exergonic reactions)
EC 7	Translocases	Transmembrane movement of ions or ionic separation (Tipton, 2018)

Table 1.4 Enzyme classes with reaction catalyzed.

Class of enzyme	Name of class	Reaction catalyzed
EC 1	Oxidoreductases	$A_{red} + B_{ox} \rightleftharpoons A_{ox} + B_{red}$
EC 2	Transferases	$A - B + C \rightarrow A + B - C$
EC 3	Hydrolases	$A - B + H_2O \rightarrow A \quad H + B - OH$
EC 4	Lyases	$A - B \rightleftharpoons A + B$ (reverse reaction : synthase)
EC 5	Isomerases	$A - B - C \rightleftharpoons A - C - B$
EC 6	Ligases	$A + B + ATP \rightarrow A - B + ADP + P_i$
EC 7	Translocases	

The first number signifies to which class the enzyme belongs to, for example, EC 1, it denotes the enzyme is oxido-reductases. EC 1 denotes that all the enzymes classified in this class will catalyze the reactions of either oxidation or reduction. Table 1.4 helps us to understand the major classes and their subclasses under the IUB system of nomenclature. The classes are further divided into subclasses that have a differentiating feature like the type of substrate or the type of product or the mechanism of a chemical reaction. On a suitable basis, the enzyme is assigned the second digit known as the subclass name. Table 1.4 provides an overview of the reactions catalyzed by each class of enzyme.

1.5 Enzymes—the present scenario

In the light of the knowledge of the past, we now look at enzymes as an efficient bioconversion tool (Toone, 2006). The role of enzymes in biological processes and their support system for cellular metabolism has been fully revealed. Presently, an upcoming field of enzymology is the intense analysis of pseudo enzymes. This branch recognizes enzymes that were catalytic in the past, but, in the process of evolution, these proteins have lost their catalytic ability. Active sites and their amino acid sequence signify their past involvement in catalysis.

Ribozymes are another type of enzyme molecule that has an immense potential to be explored currently. The enzymes are commercially used for the synthesis of industrially important products proven to have economic value due to their extreme application in normal human life. Already, enzymes are in the market proving as an excellent aid for the synthesis of commercial products like antibiotics, medicines, in bakery for hydrolysis, tenderization of meat.

1.5.1 Substrate binding

The substrate-binding site is a site that is near the active center of the enzyme where the actual catalysis takes place. The common step for initiation of catalysis reaction is that the substrate must bind to the enzyme. The specificity of binding to a particular substrate is exerted by the enzyme as an attribute of the complementary shape, the charge, the hydrophobic or hydrophilic interactions between the substrate, and the enzyme amino acids present in the catalytic in its catalytic center.

Specificity is finely demonstrated by the replicating enzymes as well as the proofreading enzymes inside the cell as they cannot afford to cause a single mistake during the process that they catalyze. The minimal error that is registered by these enzymes comes to a rate of less than 1 error in 100 million reaction polymerase. The polymerases are examples of such wonderful specificity.

The proofreading mechanism is also found in RNA polymerase, aminoacyl-t-RNA synthetases, and ribosomes but when compared, the RNA polymerases have a lesser degree of specificity as compared to other DNA polymerases.

Alternatively, some enzymes show a broad sense of specificity toward a group of substrates that are structurally related. They are active upon all the substrates that have a similar bond or similar functional group and generally are a part of a cascade of reactions.

1.5.2 Mechanism of enzyme catalyzed reaction

A normal enzyme catalyzed reaction occurs in two steps. In the initial step, the substrate binds to the enzyme and forms an intermediate that is stable and is called as an enzyme-substrate complex. The enzyme and substrate are specific, then the second part of the reaction proceeds in which the product is formed and the enzyme is liberated unaltered.

The first step is a reversible reaction and if the substrate is not specific, the enzyme and the substrate break off back into free enzyme and substrate. If the specificity is between the enzyme and the substrate that it proceeds in the forward reaction where the product is formed. The second step is an irreversible step, leading to product formation.

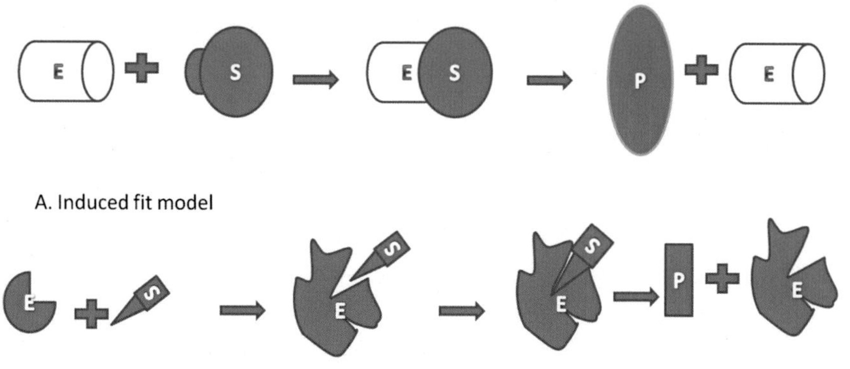

FIGURE 1.4

Models explaining the mechanism of enzyme-catalyzed reaction. (A) Lock and key model explains only the specificity of enzymes. (B) Induced fit model explains the specificity and flexibility of enzymes.

To explain the specificity of the enzyme and the mechanism of how the enzyme works, two models were proposed namely the lock and key model and induced fit model. Fig. 1.4 explains the two models proposed for explaining the key mechanism of enzyme-catalyzed reactions.

1.5.2.1 The lock and key model

The characteristics of enzymes were well explained through the lock and key model by Miller Fisher (1894). In this model, he proposed that the enzyme is a type of lock and the substrate is a type of key. The specificity that exists between the lock and the key correlates with the specificity of the enzyme and the substrate. As a particular key is meant to open for a specific lock, the same relationship exists between the enzyme and the substrate. This model was successful in explaining the specificity existing between the enzyme and the substrate but it proposed the enzyme as a rigid molecule and therefore could not have the explanation of the transition state that existed in the enzyme-catalyzed reaction.

1.5.2.2 Induced fit model

This was the second model proposed in 1958, by Daniel Koschland. This model considered enzymes as flexible structures and referred that the active site of the enzyme had the ability to change its shape due to the presence of its specific substrate or upon interaction with the specific substrate. When specific substrate approaches the enzyme molecule, it induces a structural change in the conformation of the enzyme which then favors a proper binding of the substrate with the enzyme molecule to form the enzyme-substrate complex.

The active site continuously modifies itself until and unless the substrate binds to it perfectly which is desirable for the further step of reaction. The induced fit model explains both the features of the enzyme that is the specificity and flexibility. There were experimental evidences to support the changes in the active site confirmation upon the approach of the substrate and finally getting transformed into the product. The following Fig. 1.4 explains the lock and key model in comparison to the induced fit model which gained wide acceptance.

1.5.3 Catalysis

The main principle of enzyme catalyzed reaction is its energy of activation (Ea). This quality helps the enzyme to carry out reactions at normal temperatures and pressures that otherwise would require enormous environmental conditions under chemical catalysis. The influence of lowering of energy of activation in the enzyme catalyzed reaction is the principal mechanism by which enzymes wonderfully allow reactions to occur at normal temperature and pressure at prevailing intracellular conditions.

1.6 Recent developments

The recent developments of enzymes are in direction of demand in the market as a transforming tool with reduced or low sensitivity and high rate of substrate transformation. Following are some current implications of enzyme applications.

1.6.1 Immobilization

The past 2 decades have witnessed a major focus on enzyme immobilization. The use of immobilization has been explored for improved stability and enhanced cycles of usage. Natural derivatives that organic porous materials offer an optimum surface area for immobilization. The future technologies would encompass nanotubes for better immobilization options (Neupane et al., 2019).

1.6.2 Biosensors

These are some of the best sensing devices as they get the added benefit of specificity imported due to enzyme component in the biosensors. There are two components: enzyme coupled with the signal transducer. The signals are transformed into digital or measurable outputs. Biosensors like glucometers are very primitive. Diagnostic tools used in the diagnosis of glucose, lactate, and cholesterol are enzyme-based biosensors. The common fields of usage of wireless sensors are Quality Assurance of Pasteurization of Milk, blanching of fruits and vegetables, and many more (Kaur and Kaushal, 2019).

1.6.3 Diagnostic tools

Diagnostic tools coupled with enzymes nowadays are credited for accurate and fast diagnosis. Most of such are kit-based eco-friendly and affordable by the majority of the population. Affordability has accelerated their usage and popularity in the present (Singh et al., 2019).

The future of enzymology lies in enzyme engineering. This technology has unforeseen capabilities of improvising the enzyme deficiency. The polyketides and peptides are most tried upon technologies expected to yield overwhelming results. A novel approach for enzyme modification and improved catalysis are the studies that can lead to better structure designing combined with present enzyme features. Enzyme modification and improved catalysis, computational studies can lead to a better structure designing combined with present enzyme features.

1.6.4 Metagenomics

There is an immense potential of microbes to express enzymes but a variety of them are unexplored due to the diverse cultivation requirement of microbes. Enzymes presently worked upon are derived from cultivable microbes. The researchers were trying for techniques to explore enzymes from the uncultivable (Berini et al., 2017). Metagenomics offers an option of exploring a wide variety of novel enzymes that are expressed by the diverse microbes, escaping normal isolation and cultivation techniques. Metagenomics is now used in combination with Bioinformatics. Bioinformatics has aided in exploring the structure of enzymes for multisequencing with the new enzymes and comparison of the sequence with the present database.

1.7 The future

Since past, the enzyme research has laid an immensely strong foundation and visioned the unforeseen promising horizons of enzymology. Previously, the usage of enzymes was restricted to limited industries involving bakery products as an outcome of traditional knowledge where the mechanism was still unexplored. Knowledge derived from multidisciplinary research about enzymes has revolutionized its usage in food and agriculture, medicine, and other technological by-products. The future lies in the usage of recombinant enzymes with enhanced features and product-specific approach. Metagenomics is the approach for the exploration of new enzyme options. The discovery of new enzymes would harness new possibilities of its application in exclusive enzyme reactions significant in industrial background.

The present industries have a large lacuna for catalysts with specificity, stability, and tolerance. Molecules offering all these attributes are more desirable for future industrial applications. The industries are expecting tailor-made enzymes that would be cost-effective apart from being eco-friendly.

The above industrial demand has nanocarriers as promising future enzyme technology. Nanobiocatalyst is defined as a combination of nanoparticles along with enzymes or biological catalytical elements. Stability, specificity, and tolerance are the attributes conferred due to the large surface area and small size. The future research prospects would be the immobilization of enzymes using nanomaterials. The amalgamation of nanoparticles and enzymes improves catalytic functions and productivity, which is the demand of the hour.

References

Berg, J.M., Tymoczko, J.L., Stryer, L., 2002. The Amino Acid Sequence of a Protein Determines Its Three-Dimensional Structure.Biochemistry, fifth ed. W H Freeman, New York. Section 3.6.

Berini, F., Casciello, C., Marcone, G.L., Marinelli, F., 2017. Metagenomics: novel enzymes from non-culturable microbes. FEMS Microbiol. Lett. 364 (21). https://doi.org/10.1093/femsle/fnx211.

Buchner, E., 1897. Alcoholic fermentation without yeast cells. Ber. Dtsch. Chem. Ges 30, 117–124. https://doi.org/10.1002/cber.18970300121.

Dixon, M., Webb, E.C., 1958. Enzymes, 479 (Longmans, 1958).

Florkin, M., Stotz, F. (Eds.), 1965. Comprehensive Biochemistry, vol. 13. Elsevier, Amsterdam.

Johnston, J.P., Longuet-Higgins, H.C., Ogston, A.G., 1945. Trans. Faraday Soc. 41, 588.

Kaur, K., Kaushal, P., 2019. Enzymes as Analytical Tools for the Assessment of Food Quality and Food Safety. https://doi.org/10.1016/B978-0-444-64114-4.00010-8.

Mattey, M., Simoes, D., Brown, A., Fan, X., 1998. Enzymes with a low molecular weight. Acta Chim. Slov. 45.

Mu, R., Wang, Z., Wamsley, M.C., Duke, C.N., Lii, P.H., Epley, S.E., Todd, L.C., Roberts, P.J., 2020. Application of enzymes in regioselective and stereoselective organic reactions. Catalysts 10, 832.

Neupane, S., n Patnode, K., Li, H., Baryeh, K., Liu, G., Hu, J., Chen, B., Pan, Y., Yang, Z., 2019. Enhancing enzyme immobilization on carbon nanotubes via metal–organic frameworks for large-substrate biocatalysis. ACS Appl. Mater. Interfaces 11 (12), 12133–12141. https://doi.org/10.1021/acsami.9b01077.

Nicaud, J.M., Madzak, C., van den Broek, P., Gysler, C., Duboc, P., Niederberger, P., Gaillardin, C., 2002. Protein expression and secretion in the yeast Yarrowia lipolytica. FEMS Yeast Res. 2 (3), 371–379. https://doi.org/10.1016/S1567-1356(02)00082-X. PMID: 12702287.

Singh, R., Singh, T., Singh, A., 2019. Enzymes as Diagnostic Tools. https://doi.org/10.1016/B978-0-444-64114-4.00009-1.

Somero, G.N., 1978. Temperature adaptation of enzymes: biological optimization through structure-function compromises. Annu. Rev. Ecol. Systemat. 9, 1–29.

Svedberg, T., Pedersen, K.O., 1940. The Ultracentrifuge, 406 (Oxford, 1940).

Swennen, D., Paul, M.F., Vernis, L., Beckerich, J.M., Fournier, A., Gaillardin, C., 2002. Secretion of active anti-Ras single-chain Fv antibody by the yeasts Yarrowia lipolytica and Kluyveromyces lactis. Microbiology 148 (1), 41–50. https://doi.org/10.1099/00221287-148-1-41.

Tipton, K., 2018. Translocases (EC 7): A New EC Class. Enzyme Nomenclature News.

Toone, E.J. (Ed.), 2006. Advances in Enzymology and Related Areas of Molecular Biology, Protein Evolution, vol. 75. Wiley-Interscience, ISBN 0471205036.

Wright, S., 1962. An observation suggesting that the molecular weights of enzymes can be arranged in three geometric series. Nature 193, 334–337. https://doi.org/10.1038/193334a0.

CHAPTER

Distribution and diversity in microbial enzymes

2

2.1 Origin of enzyme diversity

Since the past, the enzymes were accredited responsibility of cellular metabolism. The evolution directed by the environmental changes has bloomed the phenomena of adaptation in life forms irrespective of simple or complex. The phenotypically visible adaptations are grilled in the compromise or changes at the genetic level imparting their evolution.

Since the early period of our planet witnessed high temperatures, catalysis would have been catered by more stable and nonenzymatic molecules. These are considered to be the potential ancestors of present-day enzymes with additional features and extended catalytic capabilities.

Mutations are considered to be one of the major causes of enzyme change and its transformation with the passing of time. The rudimentary cause of the evolution of enzymes in terms of their function is genetic modification. The adaptive measure of any organism in the evolution history is to create a balance between itself and the changing environment. The change that is best suited is selected. The change may incorporate higher stability and selectivity with respect to the enzyme. This results in the broadening of diverse functions of the enzyme or the evolution of an entirely new enzyme satisfying the requirement of environmental change as an adaptive measure. This provides the grounds for the promiscuous nature of enzymes.

2.2 Natural niche as a diversity source

Our planet caters to a wide variety of environments ranging from chilled frozen zones to hot temperate areas, extreme temperature conditions in sulfur springs and high-pressure areas, and salinity regions in the marine environment. The microorganisms being ubiquitous survive even in such extreme conditions. The extremophiles confer the feature of endurance to its expressed enzymes against the harsh environment in which they survive. The diverse nature of enzymes is generated due to its environment and are called extremophiles. The extremophilic enzymes with their occurrence, environment, and microbial source are listed in Table 2.1.

Protocols and Applications in Enzymology. https://doi.org/10.1016/B978-0-323-91268-6.00013-2

Table 2.1 Extremophilic enzymes—microbial sources and their ecological niche.

Ecological niche	Enzymes expressed	Microbial occurrence	References
Acidophiles (low pH)			
Solfataric fields geothermal sulfur-rich acidic sites Mine drains Human intestine	Amylases Glucoamylases Xylanases Cellulases Proteases	*Sulfolobus solfataricus* *Sulfolobus hakonensis* *Vulcanisaeta thermophila*	Jeffries et al. (2001) Takayanagi et al. (1996) Yim et al. (2014)
Alkaliphiles (high pH)			
Deserts Hot springs Mine drains	Cellulases Amylases Lipases Proteases	*Bacillus halodurans* C-125	Fujinami and Fujisawa (2010)
Halophiles (high salt concentration)			
Salt lakes Marine coasts Artificial solar salterns Natural brines in coastal and submarine pools Deep salt mines	Lipases Amylase Endoglucanases Metalloprotease	*Halobacterium* spp. (Archaea) *Salinibacter ruber* (Bacteroidetes) *Dunaliella salina* (green alga)	DasSarma and DasSarma (2015)
Volcanic belts metal-based industries	Whole cells	*S. thermosulfidooxidans* *S. thermosulfidooxidans* *A. ferrooxidans* *Pseudomonas* *Sulfobacillus* spp. *Vibrionacea* sp. *T. thiooxidans* *T. ferrooxidans*	Shakoori et al. (2010) Rehman et al. (2009) Afrasayab et al. (2002) Lyas et al. (2007) Jadhav and Hocheng (2015)
Piezophiles/barophiles (high pressure)			
	Protease	*Methanococcus jannaschii.*	Michels and Clark (2015)
	Amylase	*Pseudomonas*-like amylase producer, strain MS300	Kobayashi et al. (1998)
	Whole organism	Proteobacteria barotolerant strain DSK25 DB5501,	Kato et al. (1995a,b) Kato et al.

Table 2.1 Extremophilic enzymes—microbial sources and their ecological niche.—*cont'd*

Ecological niche	Enzymes expressed	Microbial occurrence	References
		DB6101, DB6705, DB6906, DB172F, *Shewanella* sp. PT99 barophilic strains DSS12 and *Shewanella benthica* barophilic strains *Shewanella* sp. SC2A, Photobacterium sp. SS9 and DSJ4 *S. hanedai* and DSK1 *Shewanella* sp. SC2A, Photobacterium sp. SS9, DSJ4 *S. hanedai* DSK1	(1996a,b) Li et al. (1998)
Psychrophiles (low temperatures)			
Polar regions Deep sea or oceans	α-Amylase (10°C) Glucanases (10°C) Aminopeptidase (10°C) Carbonic anhydrase	*P. haloplanktis* *Pseudoalteromonas haloplanktis* *P. haloplanktis* *Colwellia psychrerythraea*	D'Amico et al. (2003) Chiuri et al. (2009)
Radiophiles (ionizing radiations)			
Land and sea topmost surfaces Nuclear wastes	Whole organism	*Halobacterium* sp. NRC-1 and *Deinococcus radiodurans*	DeVeaux et al. (2007)
Thermophiles (high temperature)			
Industrial wastes Deep sea and hot springs	Proteinases, Chitinases, Cellulases, Keratinases Amylases Xylanases	*Thermomonospora curvata* *Thermophillus acquaticus*	Nigam (2013) Stutzenberger (1987)
Xerophiles (scarce water)			
Deserts	β-Fructofuranosidases or invertases	*Aspergillus niger* GH1 *Actinobacteria*	Veana et al. (2014)

2.3 Evolution and function of enzymes

The study of enzyme evolution is quite interesting as a scientist has focused on similarity statistics. The similarities are on the grounds of sequence, analogy, and structural congruence. Present-day science offers innovative tools for the comparison and analysis of diverse enzyme sources with an analogy in their function. The immense research targeted toward enzymes leads to the accumulation of fast information regarding the mechanism of action, structure, and functions.

The literature cites varied examples of enzymes studied yet there was no common platform for comparison of existing enzymes. This enormous upcoming literature about enzymes generated the need of nomenclature and classification of enzymes. The preliminary or the trivial system popped up with two kinds of drawbacks represented pictorially in Fig. 2.1.

a. Multiple names to a single enzyme.
b. Similar names to two entirely different enzymes.

Therefore, a need arose for a persistent classification and nomenclature that will provide scope for newly discovered enzymes sorted in the proper group.

As already discussed, the experts of Enzyme Commission provided a wonderful nomenclature and classification system. This system classified enzymes into seven classes that were further subdivided into subclasses up to four levels. The detailed four-level subgrouping of enzymes are provided Class I in Table 2.2; Class II in Table 2.3; Class III in Table 2.4; Class IV in Table 2.5; Class V in Table 2.6; Class VI in Table 2.7; Class VII in Table 2.8. The enzyme classification and its basis are detailed in Fig. 2.2.

The most admiring feature of the classification is it takes into account the enzyme promiscuity. The enzyme evolution is an attribute acquired due to adaptive changes directed by environmental shifts.

FIGURE 2.1

The trivial system of enzyme classification and related issues.

2.4 Acidophilic enzymes

As earlier discussed in Chapter 1, acid of enzymes exhibit their optimum activity at lower pH ranging from 3.0 to 6.0. The preliminary sources of acidophilic enzymes

Table 2.2 Class 1: Oxidoreductases classification.

EC 1.1	Acting on the CH—OH group of donors
EC 1.1.1	With NAD+ or NADP+ as acceptor
EC 1.1.2	With a cytochrome as acceptor
EC 1.1.3	With oxygen as acceptor
EC 1.1.4	With a disulfide as acceptor
EC 1.1.5	With a quinone or similar compound as acceptor
EC 1.1.9	With a copper protein as acceptor
EC 1.1.98	With other, known, physiological acceptors
EC 1.1.99	With unknown physiological acceptors
EC 1.2	Acting on the aldehyde or oxo group of donors
EC 1.2.1	With NAD+ or NADP+ as acceptor
EC 1.2.2	With a cytochrome as acceptor
EC 1.2.3	With oxygen as acceptor
EC 1.2.4	With a disulfide as acceptor
EC 1.2.5	With a quinone or similar compound as acceptor
EC 1.2.7	With an iron-sulfur protein as acceptor
EC 1.2.98	With an iron-sulfur protein as acceptor
EC 1.2.99	With unknown physiological acceptors
EC 1.3	Acting on the CH—CH group of donors
EC 1.3.1	With NAD+ or NADP+ as acceptor
EC 1.3.2	With a cytochrome as acceptor
EC 1.3.3	With oxygen as acceptor
EC 1.3.4	With a disulfide as acceptor
EC 1.3.5	With a quinone or related compound as acceptor
EC 1.3.7	With an iron-sulfur protein as acceptor
EC 1.3.8	With a flavin as acceptor
EC 1.3.98	With other, known, physiological acceptors
EC 1.3.99	With unknown physiological acceptors
EC 1.4	Acting on the CH—NH_2 group of donors
EC 1.4.1	With NAD+ or NADP+ as acceptor
EC 1.4.2	With a cytochrome as acceptor
EC 1.4.3	With oxygen as acceptor
EC 1.4.4	With a disulfide as acceptor
EC 1.4.5	With a quinone or other compound as acceptor
EC 1.4.7	With an iron-sulfur protein as acceptor
EC 1.4.9	With a copper protein as acceptor
EC 1.4.98	With a copper protein as acceptor
EC 1.4.99	With unknown physiological acceptors
EC 1.5	Acting on the CH—NH group of donors
EC 1.5.1	With NAD+ or NADP+ as acceptor
EC 1.5.3	With oxygen as acceptor
EC 1.5.4	With a disulfide as acceptor
EC 1.5.5	With a quinone or similar compound acceptor

Continued

Table 2.2 Class 1: Oxidoreductases classification.—*cont'd*

EC 1.5.7	With an iron-sulfur protein as acceptor
EC 1.5.8	With a flavin or flavoprotein as acceptor
EC 1.5.98	With other, known, physiological acceptors
EC 1.5.99	With unknown physiological acceptors
EC 1.6	Acting on NADH or NADPH
EC 1.6.1	With NAD+ or NADP+ as acceptor
EC 1.6.2	With a heme protein as acceptor
EC 1.6.3	With oxygen as acceptor
EC 1.6.5	With a quinone or similar compound as acceptor
EC 1.6.6	With a nitrogenous group acceptor
EC 1.6.99	With unknown physiological
EC 1.7	Acting on other nitrogenous compounds as donors
EC 1.7.1	With NAD+ or NADP+ as acceptor
EC 1.7.2	With a cytochrome as acceptor
EC 1.7.3	With oxygen as acceptor
EC 1.7.5	With a quinone or similar compound as acceptor
EC 1.7.6	With a nitrogenous group as acceptor
EC 1.7.7	With an iron-sulfur protein as acceptor
EC 1.7.99	With unknown physiological acceptors
EC 1.8	Acting on a sulfur group of donors
EC 1.8.1	With NAD+ or NADP+ as acceptor
EC 1.8.2	With a cytochrome as acceptor
EC 1.8.3	With oxygen as acceptor
EC 1.8.4	With a disulfide as acceptor
EC 1.8.5	With a quinone or similar compound as acceptor
EC 1.8.7	With an iron-sulfur protein as acceptor
EC 1.8.98	With other, known, physiological
EC 1.8.99	With unknown physiological acceptors
EC 1.9	Acting on a heme group of donors
EC 1.9.3	With oxygen as acceptor
EC 1.9.6	With a nitrogenous group as acceptor
EC 1.9.98	With other, known, physiological acceptors
EC 1.9.99	With unknown physiological
EC 1.10	Acting on diphenols and related substances as donors
EC 1.10.1	With NAD+ or NADP+ as acceptor
EC 1.10.2	With a cytochrome as acceptor
EC 1.10.3	With oxygen as acceptor
EC 1.10.5	With a quinone or related compound
EC 1.10.9	With a copper protein as acceptor
EC 1.10.99	With unknown physiological acceptors
EC 1.11	Acting on a peroxide as acceptor
EC 1.11.1	Peroxidases
EC 1.11.2	Peroxygenases
EC 1.12	Acting on hydrogen as donor

Table 2.2 Class 1: Oxidoreductases classification.—*cont'd*

EC 1.12.1	With NAD+ or NADP+ as acceptor
EC 1.12.2	With a cytochrome as acceptor
EC 1.12.5	With a quinone or similar compound as acceptor
EC 1.12.7	With an iron-sulfur protein as acceptor
EC 1.12.98	With other, known, physiological acceptors
EC 1.12.99	With unknown physiological acceptors
EC 1.13	Acting on single donors with incorporation of molecular oxygen (oxygenases)
EC 1.13.1	Acting on single donors with incorporation of molecular oxygen (oxygenases)
EC 1.13.11	With incorporation of two atoms of oxygen
EC 1.13.12	With incorporation of one atom of oxygen (internal monooxygenases or internal mixed-function oxidases)
EC 1.13.99	Miscellaneous
EC 1.14	Acting on paired donors, with incorporation or reduction of molecular oxygen
EC 1.14.11	With 2-oxoglutarate as one donor, and incorporation of one atom of oxygen into each donor
EC 1.14.12	With NADH or NADPH as one donor, and incorporation of two atoms of oxygen into the other donor
EC 1.14.13	With NADH or NADPH as one donor, and incorporation of one atom of oxygen into the other donor
EC 1.14.14	With reduced flavin or flavoprotein as one donor, and incorporation of one atom of oxygen into the other donor
EC 1.14.15	With reduced iron-sulfur protein as one donor, and incorporation of one atom of oxygen into the other donor
EC 1.14.16	With reduced pteridine as one donor, and incorporation of one atom of oxygen into the other donor
EC 1.14.17	With reduced ascorbate as one donor, and incorporation of one atom of oxygen into the other donor
EC 1.14.18	With another compound as one donor, and incorporation of one atom of oxygen into the other donor
EC 1.14.19	With oxidation of a pair of donors resulting in the reduction of O_2 to two molecules of water
EC 1.14.20	With 2-oxoglutarate as one donor, and the other dehydrogenated
EC 1.14.21	With NADH or NADPH as one donor, and the other dehydrogenated
EC 1.14.99	Miscellaneous
EC 1.15	Acting on superoxide as acceptor
EC 1.15.1	Acting on superoxide as acceptor (only sub-subclass identified to date
EC 1.16	Oxidizing metal ions
EC 1.16.1	With NAD+ or NADP+ as acceptor
EC 1.16.3	With oxygen as acceptor
EC 1.16.5	With a quinone or similar compound as acceptor
EC 1.16.8	With a flavin as acceptor
EC 1.16.9	With a copper protein as acceptor

Continued

Table 2.2 Class 1: Oxidoreductases classification.—*cont'd*

EC 1.16.98	With other, known, physiological acceptors
EC 1.17	Acting on CH or CH_2 groups
EC 1.17.1	With NAD+ or NADP+ as acceptor
EC 1.17.2	With a cytochrome as acceptor
EC 1.17.3	With oxygen as acceptor
EC 1.17.4	With a disulfide as acceptor
EC 1.17.5	With a quinone or similar compound as acceptor
EC 1.17.7	With an iron-sulfur protein as acceptor
EC 1.17.8	With a flavin as acceptor
EC 1.17.9	With a flavin as acceptor
EC 1.17.98	With other, known, physiological acceptors
EC 1.17.99	With unknown physiological acceptors
EC 1.18	Acting on iron-sulfur proteins as donors
EC 1.18.1	With NAD+ or NADP+ as acceptor
EC 1.18.6	With dinitrogen as acceptor
EC 1.19	Acting on reduced flavodoxin as donor
EC 1.19.1	With NAD+ or NADP+ as acceptor
EC 1.19.6	With dinitrogen as acceptor
EC 1.20	Acting on phosphorus or arsenic in donors
EC 1.20.1	With NAD+ or NADP+ as acceptor
EC 1.20.2	With a cytochrome as acceptor
EC 1.20.4	With disulfide as acceptor
EC 1.20.9	With a copper protein as acceptor
EC 1.20.98	With other, known, physiological acceptors
EC 1.20.99	With unknown physiological acceptors
EC 1.21	Catalyzing the reaction X–H + Y–H = X–Y
EC 1.21.1	Catalyzing the reaction X–H + Y–H = X–Y
EC 1.21.3	With oxygen as acceptor
EC 1.21.4	With a disulfide as acceptor
EC 1.21.98	With other, known, physiological acceptors
EC 1.21.98	With other, known, physiological acceptors
EC 1.21.99	With unknown physiological acceptors
EC 1.22	Acting on halogen in donors
EC 1.22.1	With NAD+ or NADP+ as acceptor
EC 1.23	Reducing C–O–C group as acceptor
EC 1.23.1	With NADH or NADPH as donor
EC 1.23.5	With a quinone or similar compound as acceptor
EC 1.97	Other oxidoreductases
EC 1.97.1	Sole sub-subclass for oxidoreductases that do not belong in the other subclasses
EC 1.99.1	Hydroxylases (now covered by EC 1.14)
EC 1.99.2	Oxygenases (now covered by EC 1.13)

Table 2.3 Class 2: Transferases classification.

EC 2.1	Transferring one-carbon groups
EC 2.1.1	Methyltransferases
EC 2.1.2	Hydroxymethyl-, formyl- and related transferases
EC 2.1.3	Carboxy- and carbamoyltransferases
EC 2.1.4	Amidinotransferases
EC 2.1.5	Methylenetransferases
EC 2.2	Transferring aldehyde or ketonic groups
EC 2.2.1	Transketolases and transaldolases
EC 2.3	Acyltransferases
EC 2.3.1	Transferring groups other than aminoacyl groups
EC 2.3.2	Aminoacyltransferases
EC 2.3.3	Acyl groups converted into alkyl groups on transfer
EC 2.4	Glycosyltransferases
EC 2.4.1	Hexosyltransferases
EC 2.4.2	Pentosyltransferases
EC 2.4.99	Transferring other glycosyl groups
EC 2.5	Transferring alkyl or aryl groups, other than methyl groups
EC 2.5.1	Transferring alkyl or aryl groups, other than methyl groups (only sub-subclass identified to date)
EC 2.6	Transferring nitrogenous groups
EC 2.6.1	Transaminases
EC 2.6.3	Oximinotransferases
EC 2.6.99	Transferring other nitrogenous groups
EC 2.7	Transferring phosphorus containing groups
EC 2.7.1	Phosphotransferases with an alcohol group as acceptor
EC 2.7.2	Phosphotransferases with a carboxy group as acceptor
EC 2.7.2	Phosphotransferases with a carboxy group as acceptor
EC 2.7.3	Phosphotransferases with a nitrogenous group as acceptor
EC 2.7.4	Phosphotransferases with a phosphate group as acceptor
EC 2.7.6	Diphosphotransferases
EC 2.7.7	Nucleotidyltransferases
EC 2.7.8	Transferases for other substituted phosphate groups
EC 2.7.9	Phosphotransferases with paired acceptors
EC 2.7.10	Protein-tyrosine kinases
EC 2.7.11	Protein-serine/threonine kinases
EC 2.7.12	Dual-specificity kinases (those acting on Ser/Thr and Tyr residues)
EC 2.7.13	Protein-histidine kinases
EC 2.7.14	Protein-arginine kinases
EC 2.7.99	Other protein kinases
EC 2.8	Transferring sulfur-containing groups
EC 2.8.1	Sulfurtransferases
EC 2.8.2	Sulfotransferases
EC 2.8.3	CoA-transferases
EC 2.8.4	Transferring alkylthio groups
EC 2.8.5	Thiosulfotransferases
EC 2.9	Transferring selenium-containing groups
EC 2.9.1	Selenotransferases
EC 2.10	Transferring molybdenum- or tungsten-containing groups
EC 2.10.1	Molybdenumtransferases or tungsten transferases with sulfide groups as acceptors

Table 2.4 Class 3: Hydrolases classification.

EC 3.1	Acting on ester bonds
EC 3.1.1	Carboxylic-ester hydrolases
EC 3.1.2	Thioester hydrolases
EC 3.1.3	Phosphoric-monoester hydrolases
EC 3.1.4	Phosphoric-diester hydrolases
EC 3.1.5	Triphosphoric-monoester hydrolases
EC 3.1.6	Sulfuric-ester hydrolases
EC 3.1.7	Diphosphoric-monoester hydrolases
EC 3.1.8	Phosphoric-triester hydrolases
EC 3.1.11	Exodeoxyribonucleases producing 50-phosphomonoesters
EC 3.1.12	Exodeoxyribonucleases producing 30-phosphomonoesters
EC 3.1.13	Exoribonucleases producing 50-phosphomonoesters
EC 3.1.14	Exoribonucleases producing 3′-phosphomonoesters
EC 3.1.15	Exonucleases that are active with either ribo- or deoxyribonucleic acids and produce 5′-phosphomonoesters
EC 3.1.16	Exonucleases that are active with either ribo- or deoxyribonucleic acids and produce 3′-phosphomonoesters
EC 3.1.21	Endodeoxyribonucleases producing 5′-phosphomonoesters
EC 3.1.22	Endodeoxyribonucleases producing 3′-phosphomonoesters
EC 3.1.25	Site-specific endodeoxyribonucleases that are specific for altered bases
EC 3.1.26	Endoribonucleases producing 5′-phosphomonoesters
EC 3.1.27	Endoribonucleases producing 3′-phosphomonoesters
EC 3.1.30	Endoribonucleases that are active with either ribo- or deoxyribonucleic acids and produce 5′-phosphomonoesters
EC 3.1.31	Endoribonucleases that are active with either ribo- or deoxyribonucleic acids and produce 3′-phosphomonoesters
EC 3.2	Glycosylases
EC 3.2.1	Glycosidases, that is, enzymes that hydrolyze O- and S-glycosyl
EC 3.2.2	Hydrolyzing N-glycosyl compounds
EC 3.3	Acting on ether bonds
EC 3.3.1	Thioether and trialkylsulfonium hydrolases
EC 3.3.2	Ether hydrolases
EC 3.4	Acting on peptide bonds (peptidases)
EC 3.4.11	Aminopeptidases
EC 3.4.13	Dipeptidases
EC 3.4.14	Dipeptidyl-peptidases and tripeptidyl-peptidases
EC 3.4.15	Peptidyl-dipeptidases
EC 3.4.16	Serine-type carboxypeptidases
EC 3.4.17	Metallocarboxypeptidases
EC 3.4.18	Cysteine-type carboxypeptidases
EC 3.4.19	Omega peptidases
EC 3.4.21	Serine endopeptidases

Table 2.4 Class 3: Hydrolases classification.—*cont'd*

EC 3.4.22	Cysteine endopeptidases
EC 3.4.23	Aspartic endopeptidases
EC 3.4.24	Metalloendopeptidases
EC 3.4.25	Threonine endopeptidases
EC 3.4.99	Endopeptidases of unknown catalytic mechanism (sub-subclass is currently empty)
EC 3.4.99	Endopeptidases of unknown catalytic mechanism (sub-subclass is currently empty)
EC 3.5	Acting on carbon-nitrogen bonds, other than peptide bonds
EC 3.5.2	In cyclic amides
EC 3.5.3	In linear amidines
EC 3.5.4	In cyclic amidines
EC 3.5.5	In nitriles
EC 3.5.99	In other compounds
EC 3.6	Acting on acid anhydrides
EC 3.6.1	In phosphorus-containing anhydrides
EC 3.6.2	In sulfonyl-containing anhydrides
EC 3.6.3	Acting on acid anhydrides to catalyze transmembrane movement of substances
EC 3.6.4	Acting on acid anhydrides to facilitate cellular and subcellular
EC 3.6.5	Acting on GTP to facilitate cellular and subcellular movement
EC 3.7	Acting on carbon-carbon bonds
EC 3.7.1	In ketonic substances
EC 3.8	Acting on halide bonds
EC 3.8.1	In carbon-halide compounds
EC 3.9	Acting on phosphorus-nitrogen bonds
EC 3.9.1	Acting on phosphorus-nitrogen bonds (only sub-subclass identified to date
EC 3.10	Acting on sulfur-nitrogen bonds
EC 3.10.1	Acting on sulfur-nitrogen bonds (only sub-subclass identified to date)
EC 3.11	Acting on carbon-phosphorus bonds
EC 3.11.1	Acting on carbon-phosphorus bonds (only sub-subclass identified to date)
EC 3.12	Acting on sulfur-sulfur bonds
EC 3.12.1	Acting on sulfur-sulfur bonds (only sub-subclass identified to date)
EC 3.13	Acting on carbon-sulfur bonds
EC 3.13.1	Acting on carbon-sulfur bonds (only sub-subclass identified to

are archaebacteria and acidophilic bacteria and fungi (Parashar and Satyanarayana, 2018). The acidophilic property is aided with thermal stability and tolerance. Following 2.9 provides information regarding the acidophilic microbes with their enzyme potential.

Table 2.5 Class 4: Lyases classification.

EC 4.1	Carbon-carbon lyases
EC 4.1.1	Carboxy lyases
EC 4.1.2	Aldehyde lyases
EC 4.1.3	Oxo-acid lyases
EC 4.1.99	Other carbon-carbon lyases
EC 4.2	Carbon-oxygen lyases
EC 4.2.1	Hydro-lyases
EC 4.2.2	Acting on polysaccharides
EC 4.2.3	Acting on phosphates
EC 4.2.99	Other carbon-oxygen lyases
EC 4.3	Carbon-nitrogen lyases
EC 4.3.1	Ammonia lyases
EC 4.3.2	Amidine lyases
EC 4.3.3	Amine lyases
EC 4.3.99	Other carbon-nitrogen lyases
EC 4.4	Carbon-sulfur lyases
EC 4.4.1	Carbon-sulfur lyases (only sub-subclass identified to date)
EC 4.5	Carbon-halide lyases
EC 4.5.1	Carbon-halide lyases (only sub-subclass identified to date)
EC 4.6	Phosphorus-oxygen lyases
EC 4.6.1	Phosphorus-oxygen lyases (only sub-subclass identified to date)
EC 4.7	Carbon-phosphorus lyases
EC 4.7.1	Carbon-phosphorus lyases (only sub-subclass identified to date)
EC 4.99	Other lyases 173 EC 4.99.1 Sole sub-subclass for lyases that do not belong in the other subclasses

Genetic engineering is one of the tools that can work on current enzymes to improvise thermostability, acid tolerance, substrate selectivity, and other important catalytic characteristics (Joshi and Satyanarayana, 2015). The industrial relevance of acid-tolerant enzymes is immense. The commercial applications of acid and thermo-tolerant enzymes delimits its industrial usage. The need of acidophilic enzymes arises due to the extreme physicochemical environment prevalent in industrial processes requiring the catalytic function of the enzyme.

The environment that caters acidophiles are hot sulfur springs, mining regions, and even our intestine. Although the organisms exhibiting acidophilic nature tend to be thermotolerant yet they do not exhibit optimum activity at both ends. A highly thermotolerant organism will exhibit a bit lower acidophilic nature and vice versa viz. *Acidianus infernus*. The optimum temperature for growth is 90 degrees and the pH range 1.0–5.0. The most acidophilic organism Picrophilaceae grows at pH 0.7 and temperature 60°C (Segerer et al., 1986, 1988). The biggest challenge with nature that executes in acidophiles is the diverse intra- and extracellular

Table 2.6 Class 5: Isomerases classification.

EC 5.1	Racemases and epimerases
EC 5.1.1	Acting on amino acids and derivatives
EC 5.1.2	Acting on hydroxy acids and derivatives
EC 5.1.3	Acting on carbohydrates and derivatives
EC 5.1.99	Acting on other compounds
EC 5.2.1	*cis–trans* Isomerases (only sub-subclass identified to date)
EC 5.3	Intramolecular oxidoreductases
EC 5.3.1	Interconverting aldoses and ketoses, and related compounds
EC 5.3.2	Interconverting keto- and enol-groups
EC 5.3.3	Transposing C=C bonds
EC 5.3.4	Transposing S–S bonds
EC 5.3.99	Other intramolecular oxidoreductases
EC 5.4	Intramolecular transferases
EC 5.4.1	Transferring acyl groups
EC 5.4.2	Phosphotransferases (phosphomutases)
EC 5.4.3	Transferring amino groups
EC 5.4.4	Transferring hydroxy groups
EC 5.4.99	Transferring other groups
EC 5.5	Intramolecular lyases
EC 5.5.1	Intramolecular lyases (only sub-subclass identified to date)
EC 5.6	Isomerases altering macromolecular conformation
EC 5.6.1	Enzymes altering polypeptide conformation or assembly
EC 5.6.2	Enzymes altering nucleic acid conformation
EC 5.99	Other isomerases
EC 5.99.1	Sole sub-subclass for isomerases that do not belong in the other subclasses
EC 5.1	Racemases and epimerases This subclass contains enzymes that catalyze either raceme

environment. The barrier is a simple cell wall or the cell membrane. The acidophiles although maintain their life system at an acidic pH yet the intracellular functionalities are executed at a neutral range. The basic complexity lies in maintaining the transmembrane transport, high concentration of organic acids, functioning of DNA, and protein repair system. The adaptive measures at the cellular level in acidophiles are given in Fig. 2.3.

2.4.1 Adaptation at cellular level

- The membrane lipids are modified (tetra ether lipids) reduced acid hydrolysis.
- Acidophiles are organisms that are tetra-ether lipids which are more resistant to acid hydrolysis.
- Proton efflux mechanism.

Table 2.7 Class 6: Ligases classification.

EC 6.1	Forming carbon-oxygen bonds
EC 6.1.1	Ligases forming aminoacyl-tRNA and related compounds
EC 6.1.2	Acid alcohol ligases (ester synthases)
EC 6.1.3	Cyclo-ligases
EC 6.2	Forming carbon-sulfur bonds 8 EC 6.2.1 Acid-thiol ligases
EC 6.3	Forming carbon-nitrogen bonds
EC 6.3.1	Acid–ammonia (or amine) ligases (amide synthases)
EC 6.3.2	Acid–amino-acid ligases (peptide synthases)
EC 6.3.3	Cyclo-ligases
EC 6.3.4	Other carbon-nitrogen ligases
EC 6.3.5	Carbon-nitrogen ligases with glutamine as amido-N-donor
EC 6.4	Forming carbon-carbon bonds
EC 6.4.1	Ligases that form carbon-carbon bonds (only sub-subclass identified to date)
EC 6.5	Forming phosphoric-ester bonds
EC 6.5.1	Ligases that form phosphoric-ester bonds (only sub-subclass identified to date)
EC 6.6	Forming nitrogen–metal bonds 57
EC 6.6.1	Forming coordination complexes

Table 2.8 Class 7: Translocases classification.

EC 7.1	Catalyzing the translocation of hydrons
EC 7.1.1	Linked to oxidoreductase reactions
EC 7.1.2	Linked to the hydrolysis of a nucleoside triphosphate
EC 7.1.3	Linked to the hydrolysis of diphosphate
EC 7.2	Catalyzing the translocation of inorganic cations
EC 7.2.1	Linked to oxidoreductase reactions
EC 7.2.2	Linked to the hydrolysis of a nucleoside triphosphate
EC 7.2.4	Linked to decarboxylation
EC 7.3	Catalyzing the translocation of inorganic anions and their chelates
EC 7.3.2	Linked to the hydrolysis of a nucleoside triphosphate
EC 7.4	Catalyzing the translocation of amino acids and peptides
EC 7.4.2	Linked to the hydrolysis of a nucleoside triphosphate
EC 7.5	Catalyzing the translocation of carbohydrates and their derivatives
EC 7.5.2	Linked to the hydrolysis of a nucleoside triphosphate
EC 7.6	Catalyzing the translocation of other compounds 18
EC 7.6.2	Linked to the hydrolysis of a nucleoside triphosphate

Enzyme classification Nomenclature Committee of the International Union of Biochemistry and Molecular Biology (NC-IUBMB) LATEX version prepared by Andrew McDonald, School of Biochemistry and Immunology, Trinity College Dublin, Ireland Generated from the ExplorEnz database, March 2019 © 2019 IUBMB Contents.

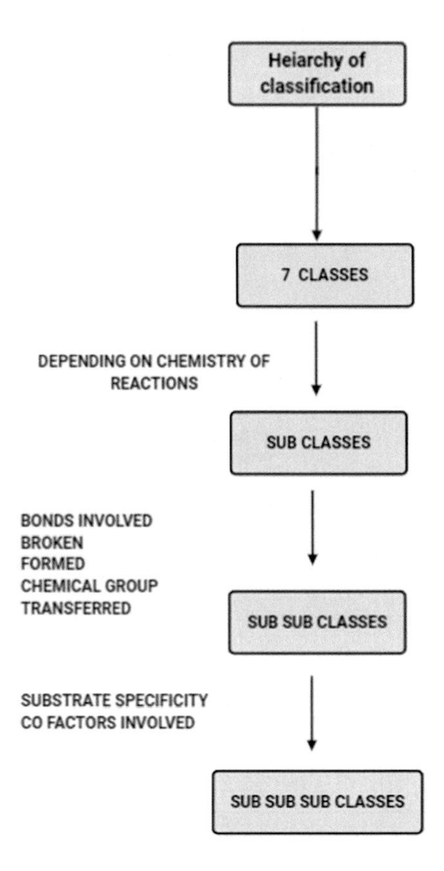

FIGURE 2.2

The four-level enzyme classification.

- ➢ Putative cation transporters involved in the generation of Donan potential.
- ➢ Active genes for acid degradation.
- ➢ Proton sequestering ability of cytoplasm.
- ➢ Hyperactive DNA repair system.
- ➢ Presence of Hsp and chaperons.
- ➢ Sugar modification on the cell surface of archeon.

Among the major classes of acidophiles, bacteria and archaea bacteria are the most dominant types prevailing in the acidic environment.

Fig. 2.4 elucidates the diversity of microbes thriving in the extreme environment of pH and temperature. As earlier stated these microbes although sustain a low

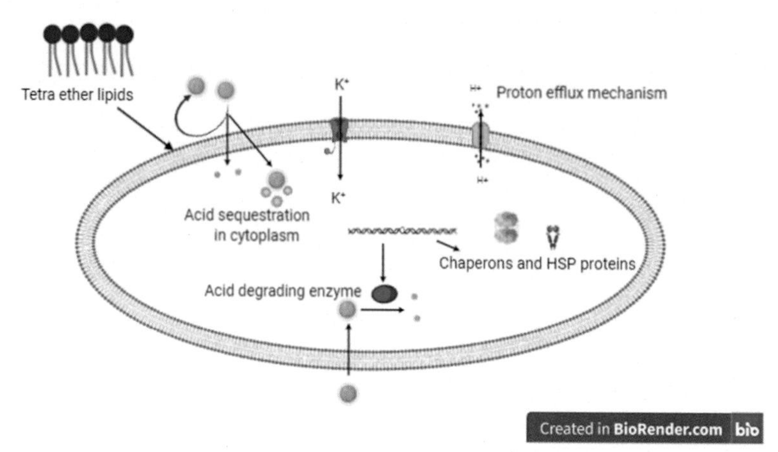

FIGURE 2.3

The adaptive measures at cellular level in acidophiles.

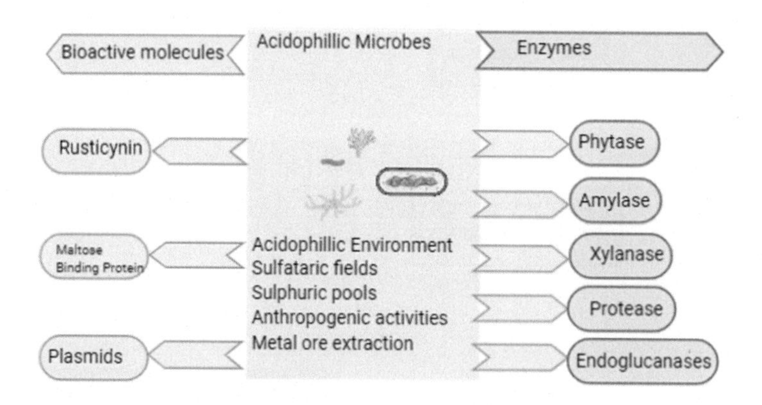

FIGURE 2.4

Acidophilic microbes—natural habitats and biomolecule resources.

environmental pH yet maintain a nearly neutral pH within them. The intracellular pH is the metabolic necessity of the organism to shield the genetic material from the deteriorative effects of ions as well as the metabolic enzymes and proteins.

The most promising enzyme with industrial potential is the process of starch hydrolysis by acidophilic amylase. The industrial parameters for the saccharification processes are optimal functioning at pH 3.0–4.5 at 70°C (Sharma and

Satyanarayana 2010). The thermostable acidophilic amylases have been reported from bacterial sources (Matzke et al., 1997; Bai et al., 2012; Sharma et al., 2016). The mode of nutrition has a directive influence on the metabolism and enzymes involved in it. The sulfur-oxidizing autotrophic bacteria is one such example. These bacteria have ability to derive electrons by oxidizing the reduced sulfur-containing compounds for their chemoautotrophic nutrition.

α-Amylase has been studied extensively for its molecular adaptation leading toward catalysis in thermoacidophilic environment. Amylase from *Bacillus acidiocola* has gained popularity as a potential thermoacidophilic enzyme. The stability of the enzyme is conferred due to surface charge density and amino acid composition. Glutamic and aspartic acid residues impart the diverse feature to this enzyme.

In comparison with the neutrophilic amylases, acidic amino acid content is found to be restricted. The comparative studies revealed that the catalytic center of the thermophilic and neutrophilic enzymes are conserved. The acid stability is therefore the virtue of the noncatalytic amino acids (Table 2.9).

Table 2.9 Acidophilic enzymes with their microbial sources.

Acidophilic enzymes and source	Optimum conditions	References
Phytase enzyme		
Thermomyces lanuginosus SSBP	55°C and pH 5.0, and was stable between 50 and 90°C from pH 3.0–6.0	Makolomakwa et al. (2017)
Thermomyces lanuginosus TL-7	2.0–10.0 and 30–90°C, respectively	Gulati et al. (2007)
Aspergillus ficuum		Coban and Demirci (2014)
Saccharomyces cerevisiae	Recombinant strain	Chen et al. (2016)
Thermomyces lanuginosus	55°C, initial pH 7.5	Berikten and Kivanc (2014)
Enterobacter sp. ACSS	(40–80°C) and pH (2.0–6.0)	Chanderman et al. (2016)
A. niger	5.0	Soni et al. (2010)
Amylases		
Bacillus sp. JAR-26	45°C at pH 5.5	Lal (2011)
Alicyclobacillus sendaiensis NUST	pH 3.5 and 85°C	Li et al. (2012)
Aspergillus niger RBP7	pH 3.0	Mukherjee et al. (2019)
A. niger and *R. stolonifera*	5.0 and 30°C	Saleem and Ebrahim (2013)

Continued

Table 2.9 Acidophilic enzymes with their microbial sources.—*cont'd*

Acidophilic enzymes and source	Optimum conditions	References
Pyrococcus sp. ST04	90–950°C and pH 5	Adrio and Demain (2014)
Haloferax mediterranei	50–60, 7.0–8.0, 10 h	Fukushima et al. (2005)
Xylanases		
A. foetidus	5.3	Shah et al. (2006)
A. awamori	5.0	Do et al. (2012)
Proteases		
Xanthomonas sp.	2.7	Oda et al. (1987a)
Pseudomonas sp.	3.0	Oda et al. (1987b)
Sulfolobus acidocaldarius	2.0	Murao et al. (1988)
Thermoplasma volcanium	3.0	Kocabıyık and Ozel (2007)
Endo-glucanases		
A. acidocaldarius	4.0	Eckert and Schneider (2003)

2.5 Alkaliphiles

Alkaliphilic microbes are well adapted to the environment bearing natural pH 9.0 or above. The enzyme produced by such organisms are extremophilic, exhibit stability as well as optimum activity at highly alkaline pH. Common enzymes are expressed as cellulases, amylases, lipases, and proteases. These enzymes harbor additional features of thermal stability and resist oxidation. *Bacillus* species has been the most common source of such enzymes. Following is the definition proposed for halophiles

> *There are no precise definitions of what characterizes an alkaliphilic or alkalitolerant organism. Several microorganisms exhibit more than one pH optimum for growth depending on the growth conditions, particularly nutrients, metal ions, and temperature ... Therefore, the term "alkaliphile" is used for microorganisms that grow optimally or very well at pH values above 9, often between 10 and 12, but cannot grow or grow only slowly at the near-neutral pH value of 6.5.*
>
> **Horikoshi (1999).**

Table 2.10 Alkaliphilic enzymes and their microbial sources.

Enzyme	Source	References
Alkaline protease	*Bacillus* sp. NPST-AK15	Ibrahim et al. (2015)
	Bacillus circulans MTCC 7942	Patil and Chaudhari (2013)
	Bacillus Staphylococcus Micrococcus Pseudomonas Arthrobacter	Kanekar et al. (2002)
α-Amylases	*Bacillus licheniformis* Isolate AI20	Yasser et al. (2013)
	Bacillus subtilis	Nuha et al. (2016)
Cyclodextrin	*Microbacterium terrae* KNR 9	Rajput et al. (2016)
Glucotransferase	*Bacillus pseudalcaliphilus* 20RF	Atanasova et al. (2011)
	Bacillus Strains	Atanasova et al. (2008)
Cellulases	*Aspergillus* sp. M-HP32, *Penicillium* sp.LM-HP33, *Penicillium* sp. LM-HP37	Vega et al. (2012)
Lipases	*Geobacillus thermoleovorans* DA2	Fotouh et al. (2016)
	Archaeoglobus fulgidus	Chen et al. (2009)
Hydrocarbon degrading enzymes	*Alcanivorax borkumensis*	Kadri et al. (2018)
Xylanases	*Bacillus licheniformis* *Bacillus* sp. *Arthrobacter* sp. MTCC 5214 *Aspergillus terreus* UL *Streptomyces cyaneus* SN32	Raj and Suman (2010) Kamble and Jadhav (2012) Khandeparkar and Bhosle (2006) Chidi et al. (2008) Ninawe et al. (2008)

Alkaline environments have been a source of human attractions since the past. The alkaline lakes with thought to be devoid of life. In contrast, the alkaline environment is now revealing a great deal of microbial diversity. Table 2.10 provides an insight into the microbial sources of the alkaliphilic enzymes.

The alkaliphiles also face a similar challenge of maintaining a neutral pH in an alkaline environment. The pH of the cytoplasm should not be acidic. The proteins and exposed part of the cell membrane are alkali-resistant and adapted for ATP synthesis. The organisms that can survive in these extreme conditions of pH are eukarya, archaebacteria and eubacteria.

At the cellular level, the alkaliphilic bacteria maintain a balance between the intracellular and extracellular diversities by cellular adaptations. Some common cellular adaptations are in form of active and passive methods of regulation:

- The passive methods of regulation are the presence of polyamines in the cytoplasm.
- Reduced membrane permeability.

- Sodium channels have a role in the active regulation.
- Specific amino acid composition for salt expose proteins.
- Excessive synthesis of osmolites by the cell of the membrane.
- Membrane permeability is highly reduced.
- The cell is adapted to manage high oxidative stress.
- Sodium ion channels play an active role in homeostasis at the cellular level.
- Salt-dependent adjustments are found for plasma membrane fluidity.

Fig. 2.5 helps to understand the common cellular adaptation for maintaining homeostasis and an alkaline environment.

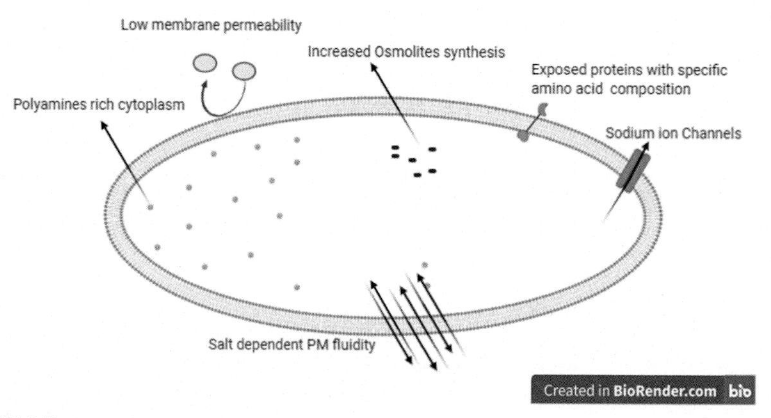

FIGURE 2.5

Strategies of homeostasis in alkaliphiles.

2.6 Halophilic organisms

Intensive research has been carried out in bacteria, archae, fungi, and green algae survivors under halophilic environment. Protozoa are recently screened for their homeostasis in an alkaline environment (Harding et al., 2017). The genetic studies revealed hyperactivity of genes related to chaperons, sodium/hydrogen transporters, lipid transport, and synthesis of sterols glycosidase are some common measures in such extremophiles.

The challenges faced by halophiles are in expressed oxidative stress, ionic stress, dehydration conditions, and reduced solubility. Common observations in halophiles are to compete with high ionic concentrations, if the environment is salty. The exposed proteins exhibits specific amino acid composition (Paul et al., 2008), synthesis of osmolites (Borowitzka and Brown, 1974). The membrane fluidity is controlled at high salinity by keeping the phospholipid ratio less, decreasing the fatty acid link and saturation of the level of phospholipids (Turk et al., 2004, 2007). Studies in the plants have guided that salt tolerance is a consequence that helps in the automatic management of oxidative stress (Hernandez et al., 1995). Such conditions influence the mitochondria and even causes a reverse electron transport. Studies conducted on yeast *Saccharomyces cerevisiae* at salt concentrations of 3% indicated that the chaperons and hsp70 and 90 help in regulating the salt stress and controlling the protein damage under the high salt stress condition (Petrovic, 2006; Vaupotic and Plemenitas, 2007). Yet another type of organism observed in the halophilic environment is protists that have attracted attention as a source of halophilic enzymes (Foissner et al., 2014).

The marine environment is a region of exemplary biodiversity. Marine fungi are exuberant sources of degradative enzymes. As they play a vital role in nutrient recycling occurring in oceans, the enzymes from marine fungi exhibit functionality in a high degree of pH variation and temperature. Some enzymes demonstrate unparalleled ability to catalyze reactions at alkaline pH and low temperature. The recycling ability of marine microbes is extremely important in maintaining the ecological balance.

The marine fungus is an umbrella term used for fungi that are isolated from the marine environment. In many instances, it is not possible to strictly classify them as facultative and obligate. The occurrence of these fungi is reported to be associated with the availability of substrate. Due to the environmental factors, terrestrial fungi are introduced in the marine environment and generate great diversity. This marked the beginning of the research diverted toward economic products derived from these fungi. The attributes of these fungi are found to be quite different, probably due to the impact of extreme conditions such as salinity, high pressure, temperature, oligotrophic conditions, extremes in pH, and high concentration of minerals.

The physiological performance of such marine fungi is entirely different from their terrestrial homologous counterpart. The enzymes derived from the marine-derived fungi found to have potential applications in almost all industrial fields such as food, fodder, beverage, environmental remediation, and pharmaceuticals.

The genetic sources of these highly potent enzymes are transferred through genetic engineering to facilitate their bulk production. The metagenomics can be a future tool for the production of such enzymes. Halophilic enzymes and their sources are enlisted in Table 2.11.

Table 2.11 Halophilic enzymes and their sources.

Hallophilic enzymes	Source	References
D-Mannonate dehydratase	*Thermoplasma acidophilum*	Kopp et al. (2019)
Rnase H1	*Halobacterium*	Ohtani et al. (2004)
Alkaline metalloprotease	*Salinivibrio* sp. strain AF-2004	Amoozegar et al. (2007)
Thermohalophilic lipase	*Salinivibrio* sp. strain SA-2	Amoozegar et al. (2008)
Endoglucanase	*Bacillus* sp. C14	Aygan and Arikan (2008a)
α-Amylase	*Bacillus* sp. AB68 *Streptomyces* sp D1 *Halothermothrix orenii* archaeon *Natronococcus* sp. strain Ah-36 *Natronococcus* sp. strain Ah-36	Aygan and Arikan (2008b) Chakraborty et al. (2009) Tan et al. (2008); Raj and Suman (2010) Kobayashi et al. (1994)
Protease	*Nesterenkonia* sp. AL20 archaebacterium strain 172P1 *Geomicrobium* sp. EMB2 (MTCC 10310 *Halobacillus karajensis* *Virgibacillus* sp. SK33 *Bacillus* sp. EMB9 *Salicola* sp. IC10 *Chaetomium thermophilum*	Bakhtiar et al. (2005); Kamekura and Seno (1990) Karan et al. (2011) Karbalaei-Heidari et al. (2009) Sinsuwan et al. (2010) Sinha and Khare (2014) Moreno et al. (2009) Li and Li (2009)
Cellulase	*Halocella Cellulolytica*	Bolobova et al. (1992)
Lipase	*Natronococcus* sp. (Archeaon)	Boutaiba et al. (2006)
Glucose dehydrogenase	*Haloferax mediterranei*	Britton et al. (2006)
Xylanase	*Gracibacillus* sp. TSCPVG *Glaciecola mesophila*	Giridhar and Chandra (2010) Guo et al. (2009)
Hydrolase	*Nesterenkonia* sp.	Govender et al. (2009)
Chitinase	*Halobacterium salinarum*	Hatori et al. (2006)
Serine proteinase	*Filobacillus* sp. RF2-5	Hiraga et al. (2005)
Nuclease H	*Micrococcus varians* sp.	Kamekura and Onishi (1978)

2.7 The protist—a novel enzyme source

This heterotrophic fungus such as protist is exuberant source of enzymes. They are known to express lucrative enzymes. There are 10 genera reported with 40 species and to express economically important and valuable products. The common enzymes that are produced by these protists are degradative enzymes for carbohydrates, proteins, and lipids (Kumavath, 2011). It is suggested that the marine protist has an important role in nutrient recycling in the marine environment similar to the marine fungi. Fig. 2.6 explains the role of protists in ecology.

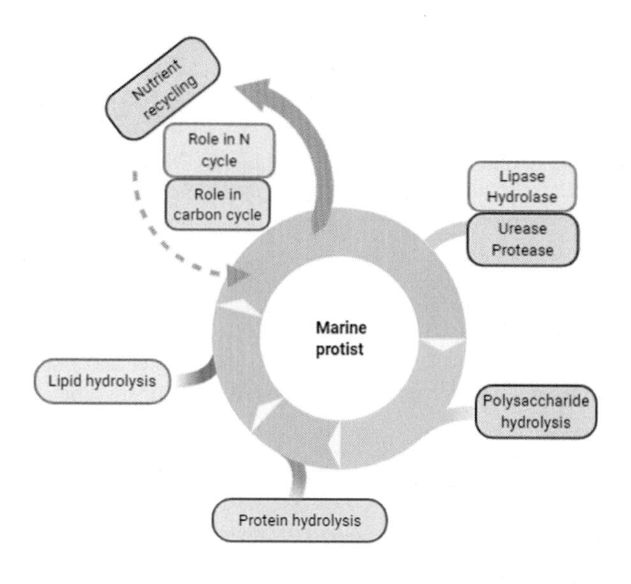

FIGURE 2.6

The role of protists in marine ecology.

2.8 Future directions

Extremophiles and their metabolic chemistry unleash the basic living mechanism and adaptive measures for their survival. The present studies are focused on sustainability at a single front. In the future to design industrially relevant enzymes, multifaceted extremophilic microbes are to be experimented as the source. The extremophilic enzymes have opened a new vision toward the life processes and their extent of occurrence under absurd conditions. However, these studies require an evolutionary linkage to be established in relation to structure and function. Modern screening techniques are being scrutinized for successful accomplishment of desired enzyme selection discussed in the forthcoming chapter.

References

Adrio, J.L., Demain, A.L., 2014. Microbial enzymes: tools for biotechnological processes. Biomolecules 16, 117–139.

Afrasayab, S., Yasmin, A., Hasnain, S., 2002. Characterization of some indigenous mercury resistant bacteria from polluted environment. Pak. J. Biol. Sci. 5, 792–797.

Amoozegar, M.A., Fatemi, Z.A., Karbalaei-Heidari, H.R., Razavi, M.R., 2007. Production of an extracellular alkaline metalloprotease from a newly isolated, moderately halophile, *Salinivibrio* sp. strain AF-2004. Microbiol. Res. 162, 369–377.

Amoozegar, M.A., Salehghamari, E., Khajeh, K., Kabiri, M., Naddaf, S., 2008. Production of an extracellular thermohalophilic lipase from a moderately halophilic bacterium, *Salinivibrio* sp. strain SA-2. J. Basic Microbiol. 48, 160–167 (CrossRefPubMedGoogle Scholar).

Atanasova, N., Petrova, P., Ivanova, V., Yankov, D., Dragomir, V., Vassileva, A., Tonkova, 2008. Isolation of novel alkaliphilic *Bacillus* strains for cyclodextrin glucanotransferase production. Appl. Biochem. Biotechnol. 149, 155–167. https://doi.org/10.1007/s12010-007-8128-5.

Atanasova, N., Kitayska, T., Bojadjieva, I., Yankov, D., Tonkova, A., 2011. A novel cyclodextrin glucanotransferase from alkaliphilic Bacillus pseudalcaliphilus 20RF: purification and properties. Process Biochem. 46 (1), 116–122.

Aygan, A., Arikan, B., 2008. A new haloalkaliphilic, thermostable endoglucanse from moderately halophilic *Bacillus* sp. C14 isolated from Van Soda Lake. Int. J. Agric. Biol. 10, 369–374.

Aygan, A., Arikan, B., Korkmas, H., Dincer, S., Colak, O., 2008. Highly thermostable and alkaline α- amylase from a halotolerant alkaliphilic *Bacillus* sp. AB68. Braz. J. Microbiol. 39, 547–553.

Bai, Y., Huang, H., Meng, K., Shi, P., Yang, P., Luo, H., Luo, C., Feng, Y., Zhang, W., Yao, B., 2012. Identification of an acidic α-amylase from *Alicyclobacillus* sp. A4 and assessment of its application in the starch industry. Food Chem. 131, 1473–1478.

Bakhtiar, S., Estiveira, R.J., Hatti-Kaul, R., 2005. Substrate specificity of alkaline protease from alkaliphilic feather-degrading *Nesterenkonia* sp. AL20. Enzym. Microb. Technol. 37, 534–540.

Berikten, D., Kivanc, M., 2014. Optimization of solid-state fermentation for phytase production by Thermomyces lanuginosus using response surface methodology. Prep. Biochem. Biotechnol. 144 (8), 834–848. https://doi.org/10.1080/10826068.2013.868357. PMID: 24279930.

Bolobova, A.V., Simankova, M.V., Markovitch, N.A., 1992. Cellulase complex of a new halophilic bacterium, *Halocella Cellulolytica*. Microbiology 61, 557–562.

Borowitzka, L.J., Brown, A.D., 1974. Salt relations of marine and halophilic species of unicellular green alga Dunaliella: role of glycerol as a compatible solute. Arch. Microbiol. 96, 37–52. https://doi.org/10.1007/BF00590161.

Boutaiba, S., Bhatnagar, T., Hacene, H., Mitchell, D.A., Baratti, J.C., 2006. Preliminary characterization of a lipolytic activity from an extremely halophilic archaeon, *Natronococcus* sp. J. Mol. Catal. B Enzyme 41, 21–26 (CrossRefGoogle Scholar).

Britton, K.L., Baker, P.J., Fisher, M., Ruzheinikov, S., Gilmour, D.J., Bonete, M.J., et al., 2006. Analysis of protein solvent interactions in glucose dehydrogenase from the extreme halophile *Haloferax mediterranei*. Proc. Natl. Acad. Sci. U.S.A. 103, 4846–4851.

Chakraborty, S., Khopade, A., Kokare, C., Mahadika, K., Chopade, B., 2009. Isolation and characterization of novel α-amylase from marine *Streptomyces* sp. D1. J. Mol. Catal. B Enzyme 58, 17–23.

Chanderman, A., Puri, A.K., Permaul, K., Singh, S., 2016. Production, characteristics and applications of phytase from a rhizosphere isolated *Enterobacter* sp. ACSS. Bioprocess. Biosyst. Eng. 39 (10), 1577–1587. https://doi.org/10.1007/s00449-016-1632-7. Epub 2016 Jun 1.PMID: 27250653.

Chen, C.K., Lee, G.C., Ko, T.P., Guo, R.T., Huang, L.M., Liu, H.J., Ho, Y.F., Shaw, J.F., Wang, A.H., 2009. Structure of the alkalohyperthermophilic *Archaeoglobusfulgidus* lipase contains a unique C-terminal domain essential for long-chain substrate binding. J. Mol. Biol. 390, 672–685.

Chen, X., Xiao, Y., Shen, W., Govender, A., Zhang, L., Fan, Y., Wang, Z., 2016. Display of phytase on the cell surface of *Saccharomyces cerevisiae* to degrade phytate phosphorus and improve bioethanol production. Appl. Microbiol. Biotechnol. 100 (5), 2449–2458. https://doi.org/10.1007/s00253-015-7170-4. Epub 2015 Nov 26.PMID: 26610799.

Chidi, B., Godana, B., Ncube, I., Van Rensburg, E.J., Cronshaw, A., Abotsi, E.K., 2008. Production, purification and characterization of celullase-free xylanase from *Aspergillus terreus* UL 4209. Afr. J. Biotechnol. 7 (21), 3939–3948.

Chiuri, R., Maiorano, G., Rizzello, A., del Mercato, L.L., Cingolani, R., Rinaldi, R., Maffia, M., Pompa, P.P., 2009. Exploring local flexibility/rigidity in psychrophilic and mesophilic carbonic anhydrases. Biophys. J. 96 (4), 1586–1596.

Coban, H.B., Demirci, A., 2014. Enhanced submerged *Aspergillus ficuum* phytase production by implementation of fed-batch fermentation. Bioprocess. Biosyst. Eng. 37 (12), 2579–2586. https://doi.org/10.1007/s00449-014-1236-z. Epub 2014 Jun 24.PMID: 24958522.

D'Amico, S., Marx, J.C., Gerday, C., Feller, G., 2003. Activity - stability relationships in extremophilic enzymes. J. Biol. Chem. 278, 7891–7896.

DasSarma, S., DasSarma, P., 2015. Halophiles and their enzymes: negativity put to good use. Curr. Opin. Microbiol. 25, 120–126.

DeVeaux, L.C., Muller, J.A., Smith, J., Petrisco, J., Wells, D.P., DasSarma, S., 2007. Extremely radiation-resistant mutants of a halophilic archaeon with increased single-stranded DNA-binding protein (RPA) gene expression. Radiat. Res. 168 (4), 507–514.

Do, T.T., Quyen, D.T., Dam, T.H., 2012. Purifification and characterization of an acid-stable and organic solvent-tolerant xylanase from *Aspergillus awamori* VTCC-F312. Sci. Asia 38, 157–165.

Eckert, K., Schneider, E., 2003. A thermoacidophilic endoglucanase (CelB) from A*licyclobacillus acidocaldarius* displays high sequence similarity to arabinofuranosidases belonging to family 51 of glycoside hydrolases. Electron. J. Biotechnol. 270, 3593–3602.

Foissner, W., Jung, J.H., Filker, S., Rudolph, J., Stoeck, T., 2014. Morphology, ontogenesis and molecular phylogeny of *Platynematum salinarum* nov. spec., a new scuticociliate (Ciliophora, Scuticociliatia) from a solar saltern. Eur. J. Protistol. 50, 174–184. https://doi.org/10.1016/j.ejop.2013.10.001.

Fotouh, A.D.M., Bayoumi, R.A., Mohamed Hassan, A., 2016. Production of thermoalkaliphilic lipase from *Geobacillus thermoleovorans* DA2 and application in leather industry. Enzym. Res. 1–9. https://doi.org/10.1155/2016/9034364, 2016.

Fujinami, S., Fujisawa, M., 2010. Industrial applications of alkaliphiles and their enzymes – past, present and future. Environ. Technol. 31 (8–9), 845–856. https://doi.org/10.1080/09593331003762807.

Fukushima, T., Mizuki, T., Echigo, A., Inoue, A., Usami, R., 2005. Organic solvent tolerance of halophilic α-amylase from a haloarchaeon, *Haloarcula* sp.strain S-1. Extremophiles 9, 85–89.

Giridhar, P.V., Chandra, T.S., 2010. Production of novel haloalkali- thermostable xylanase by a newly isolated moderately halophilic and alkali tolerant *Gracibacillus* sp. TSCPVG. Process Biochem. 45, 1730–1737.

Govender, L., Naidoo, L., Setati, M.E., 2009. Isolation of hydrolase producing bacteria from Sua pan solar salterns and the production of endo-1,4-β-xylanase from a newly isolated haloalkaliphilic *Nesterenkonia sp*. Afr. J. Biotechnol. 8, 5458–5466.

Gulati, H.K., Chadha, B.S., Saini, H.S., 2007. Production, purification and characterization of thermostable phytase from thermophilic fungus Thermomyces lanuginosus TL-7. Acta Microbiol. Immunol. Hungun. 54 (2), 121–138. https://doi.org/10.1556/AMicr.54.2007.2.3.

Guo, B., Chen, X.L., Sun, C.Y., Zhou, B.C., Zhang, Y.Z., 2009. Gene cloning, expression and characterization of a new cold-active and salt-tolerant endo-β-xylanase from marine *Glaciecola mesophila* KMM 241. Appl. Microbiol. Biotechnol. 84, 1107–1115.

Harding, T., Roger, A.J., Simpson, A.G.B., 2017. Adaptations to high salt in a halophilic protist: differential expression and gene acquisitions through duplications and gene transfers. Front. Microbiol. 8, 944. https://doi.org/10.3389/fmicb.2017.00944.

Hatori, Y., Sato, M., Orishimo, K., Yatsunami, R., Endo, K., Fukui, T., Nakamura, S., 2006. Characterization of recombinant family 18 chitinase from extremely halophilic archaeon *Halobacterium salinarum* strain NRC-1. Chitin. Chitosan. Res. 12, 201 (Google Scholar).

Hernandez, J.A., Olmos, E., Corpas, F.J., Sevilla, F., Del Río, L.A., 1995. Salt induced oxidative stress in chloroplasts of pea plants. Plant Sci. 105, 151–167. https://doi.org/10.1016/0168-9452(94)04047-8.

Hiraga, K., Nishikata, Y., Namwong, S., Tanasupawat, S., Takada, K., Oda, K., 2005. Purification and characterization of serine proteinase from a halophilic bacterium, *Filobacillus* sp.RF2-5. Biosci. Biotechnol. Biochem. 69, 38–44 (CrossRefPubMedGoogle Scholar).

Horikoshi, K., 1999. Alkaliphiles: some applications of their products for biotechnology. Microbiol. Mol. Biol. Rev. 63 (4), 735–750.

Ibrahim, A.S.S., Al Salamah, A.A.A., Elbadawi, Y.B., El- Tayab, M.A., Ibrahim, S.S.S., 2015. Production of extracellular alkaline protease by new halotolerant alkaliphilic Bacillus sp. NPST-AK15 isolated from hyper saline soda lakes. Electron. J. Biotechnol. 18 (3). https://doi.org/10.1016/j.ejbt.2015.04.001.

Jadhav, U., Hocheng, H., 2015. Enzymatic bioleaching of metals from printed circuit board. Clean Technol. Environ. Policy 17, 947–956. https://doi.org/10.1007/s10098-014-0847-z.

Jeffries, A.C., Kozera, C.J., Medina, N., Peng, X., Thi-Ngoc, H.P., Redder, P., Schenk, M.E., Theriault, C., Tolstrup, N., Charlebois, R.L., Doolittle, W.F., Duguet, M., Gaasterland, T., Garrett, R.A., Ragan, M.A., Sensen, C.W., Vander Oost, J., 2001. The complete genome of the Crenarchaeon *Sulfolobus solfataricus*P2. Proc. Natl. Acad. Sci. U.S.A. 98, 7835–7840.

Joshi, S., Satyanarayana, T., 2015. In vitro engineering of microbial enzymes with multifarious applications: prospects and perspectives. Bioresour. Technol. 176, 273–283.

Kadri, T., Rouissi, T., Magdouli, S., Brar, S.K., Hegde, K., Khiari, Z., Daghrir, R., Lauzon, J.M., 2018. Production and characterization of novel hydrocarbon degrading enzymes from *Alcanivorax borkumensis*. Int. J. Biol. Macromol. 112, 230–240.

Kamble, R.D., Jadhav, A.R., 2012. Isolation, Purification and characterization of xylanase produced by a new species of Bacillus in Solid state fermentation. Int. J. Microbiol. Article ID 683193.

Kamekura, M., Onishi, H., 1978. Properties of the halophilic nuclease of a moderate halophile, Micrococcus varians subsp. halophilus. J. Bacteriol. 133 (1), 59–65.

Kamekura, M., Seno, Y., 1990. A halophilic extracellular protease from a halophilic archaebacterium strain 172P1. Biochem. Cell. Biol. 68, 352–359 (CrossRefPubMedGoogle Scholar).

Kanekar, P.P., Nilegaonkar, S.S., Sarnaik, S.S., Kelkar, A.S., 2002. Optimization of protease activity of alkaliphilic bacteria isolated from an alkaline lake in India. Bioresour. Technol. 85 (1), 87–93. https://doi.org/10.1016/s0960-8524(02)00018-4. PMID: 12146649.

Karan, R., Singh, S., Kapoor, S., Khare, S., 2011. A novel organic solvent tolerant protease from a newly isolated *Geomicrobium* sp. EMB2 (MTCC 10310): production optimization by response surface methodology. New Biotechnol. 28, 136–145 (CrossRefGoogle Scholar).

Karbalaei-Heidari, H.R., Amoozegar, M.A., Hajighasemi, M., Ziaee, A.A., Ventosa, A., 2009. Production, optimization and purification of a novel extracellular protease from the moderately halophilic bacterium *Halobacillus karajensis*. J. Ind. Microbiol. Biotechnol. 36, 21–27.

Kato, C., Suzuki, S., Hata, S., Ito, T., Horikoshi, K., 1995a. The properties of a protease activated by high pressure from Sporosarcina sp. strain DSK25 isolated from deep-sea sediment. JAMSTEC R. 32, 7–13.

Kato, C., Sato, T., Horikoshi, K., 1995b. Isolation and properties of barophilic and barotolerant bacteria from deep-sea mud samples. Biodiv. Conserv. 4, 1–9.

Kato, C., Masui, N., Horikoshi, K., 1996a. Properties of obligatory barophilic bacteria isolated from a sample of deep-sea sediment from the Izu-Bonin trench. I Mar. Biotechnol. 4, 96–99.

Kato, C., Inoue, A., Horikoshi, K., 1996b. Isolating and characterizing deep sea marine microorganisms. Trends Biotechnol. 14, 6–12.

Khandeparkar, D.S., Bhosle, N.B., 2006. Isolation, purification and characterization of the xylanase produced by *Arthrobacter sp.* MTCC 5214 when grown in solid-state fermentation. Enzym. Microb. Technol. 39 (4), 732–742.

Kobayashi, T., Kanai, H., Aono, R., Horikoshi, K., Kudo, T., 1994. Cloning, expression and nucleotide sequencing of α- amylase gene from the haloalkaliphilic archaeon Natronococcus sp. strain Ah-36. J. Bacteriol. 176, 5131–5134.

Kobayashi, H., Takaki, Y., Kobata, K., Takami, H., Inoue, A., 1998. Characterization of α-maltotetraohydrolase produced by *Pseudomonas* sp. MS300 isolated from the deepest site of the Mariana Trench. Extremophiles 2, 401–407. https://doi.org/10.1007/s007920050085.

Kocabıyık, S., Ozel, H., 2007. An extracellular—pepstatin insensitive acid protease produced by Thermoplasma volcanium. Bioresour. Technol. 98 (1), 112–117.

Kopp, D., Willows, R., Sunna, A., 2019. Characterisation of the first archaeal mannonate dehydratase from *Thermoplasma acidophilum* and its potential role in the catabolism of D-mannose. Catalysts 9, 234. https://doi.org/10.3390/catal9030234.

Kumavath, R., 2011. Microbial production of novel enzymes and their biological applications. Adv. Microbiol. https://doi.org/10.4236/aim.2012.

Lal, N., 2011. Production of thermostable and acidophilic amylase from thermophilic Bacillus licheniformis JAR-26. J. Appl. Biol. Sci. 3, 7–12.

Li, A.N., Li, D.C., 2009. Cloning, expression and characterization of the serine protease gene from *Chaetomium thermophilum*. J. Appl. Microbiol. 106, 369–380.

Li, L., Kato, C., Horikoshi, K., 1998. Distribution of the pressure regulated • - operons in deep-sea bacteria. FEMS Microbiol. Lett. 159, 159–166.

Li, D., Sheng, J., Yang, J., Yang, C., Zeng, Z., Sheng, L., 2012. Characterization of an acidophilic and thermostable α-amylase from A*licyclobacillus sendaiensis* NUST. Starch 64 (11), 914–920.

lyas, S., Anwar, M.A., Niazi, S.B., Ghauri, M.A., 2007. Bioleaching of metals from electronic scrap by moderately thermophilic acidophilic bacteria. Hydrometallurgy 88, 180–188.

Makolomakwa, M., Puri, A.K., Permaul, K., Singh, S., 2017. Thermo-acid-stable phytase-mediated enhancement of bioethanol production using *Colocasia esculenta*. Bioresour. Technol. 235, 396–404.

Matzke, J., Schwermann, B., Baker, E.P., 1997. Acidstable and acidophilic proteins: the example of the alpha amylase from *Alicyclobacillus acidocaldarius*. Comp. Biochem. Physiol. Mol. Integr. Physiol. 118, 411–419.

Michels, P.C., Clark, D.S., 2015. Pressure-enhanced activity and stability of a hyperthermophilic protease from a deep-sea methanogen. Appl. Environ. Microbio. 63, 3985–3991.

Moreno, M.L., García, M.T., Ventosa, A., Mellado, E., 2009. Characterization of *Salicola* sp. IC10, a lipase- and protease-producing extreme halophile. FEMS Microbiol. Ecol. 68, 59–71 (CrossRefGoogle Scholar).

Mukherjee, R., Paul, T., Soren, J.P., Halder, S.K., Mondal, K.C., Pati, B.R., Mohapatra, P.K.D., 2019. Acidophilic α-amylase production from *Aspergillus niger* RBP7 using potato peel as substrate: a waste to value added approach. Waste Biomass Valor. 10, 851–863. https://doi.org/10.1007/s12649-017-0114-8.

Murao, S., Okhuni, K., Naganao, M., 1988. A novel thermostable S-PI (pepstatin Ac)-insensitive acid proteinase from thermophilic *Bacillus* novo sp. strain Mn-32. Agric. Biol. Chem. 52, 1029–1031.

Nigam, P.S.N., 2013. Microbial enzymes with special characteristics for biotechnological applications. Biomolecules 3, 597–611.

Ninawe, S., Kapoor, M., Kuhad, R.C., 2008. Purification and characterization of extracellular xylanase from *Streptomyces yaneus* SN32. Bioresour. Technol. 99 (5), 1252–1258.

Nuha, B.A.J., Madeha, N.A., Youssri, M.A., 2016. Optimization of alkaline α-amylase production by thermophilic *Bacillus subtilis*. Afr. J. Tradit. Complem. Altern. Med. 14 (1), 288–301. https://doi.org/10.21010/ajtcam.v14i1.31.

Oda, K., Nakazima, T., Terashita, T., Suziki, K.A., Murao, S., 1987a. Purifification and properties of an S-PI (Pepstatin Ac) insensitive carboxyl proteinase from a *Xanthomonas* sp. bacterium. Agric. Biol. Chem. 51, 3073–3080.

Oda, K., Sugitani, M., Fukuhara, K., Murao, S., 1987b. Purifification and properties of a pepstatin insensitive carboxyl proteinase from a gram negative bacterium. Biochim. Biophys. Acta 923, 463–469.

Ohtani, N., Yanagawa, H., Tomita, M., Itaya, M., 2004. Identification of the first archaeal Type 1 RNase H gene from Halobacterium sp. NRC-1: archaeal RNase HI can cleave an RNA-DNA junction. Biochem. J. 1 (381), 795–802. https://doi.org/10.1042/BJ20040153. PMID: 15115438; PMCID: PMC1133889.

Parashar, D., Satyanarayana, T., 2018. An insight into ameliorating production, catalytic efficiency, thermostability and starch saccharification of acid-stable α-amylases from acidophiles. Front. Bioeng. Biotechnol. 6. https://doi.org/10.3389/fbioe.2018.00125.

Patil, U., Chaudhari, A., 2013. Production of Alkaline protease by solvent tolerant Alkaliphilic *Bacillus circulans* MTCC 7942 isolated from Hydrocarbon contaminated Habitat: process parameters optimization. Int. Scholar. Res Notices. https://doi.org/10.1155/2013/942590. Article ID 942590.

Paul, S., Bag, S.K., Das, S., Harvill, E.T., Dutta, C., 2008. Molecular signature of hypersaline adaptation: insights from genome and proteome composition of halophilic prokaryotes. Genome Biol. 9, R70. https://doi.org/10.1186/gb-2008-9-4-r70.

Petrovic, U., 2006. Role of oxidative stress in the extremely salt-tolerant yeast Hortaea werneckii. FEMS Yeast Res. 6, 816–822. https://doi.org/10.1111/j.1567-1364.2006.00063.x.

Raj, E., Suman, C.E., 2010. Purification and characterization of a new hyperthermostable, allosamidin-insensitive and denaturation-resistant chitinase from the hyper thermophilic archaeon *Thermococcus chitonophagus*. Extremophiles 7, 43–53.

Rajput, K.N., Patel, K.C., Trivedi, U.B.A., 2016. Novel cyclodextrin glucanotransferase from an alkaliphile *Microbacterium terrae* KNR purification and properties. 3 Biotech 6, 168. https://doi.org/10.1007/s13205-016-0495-6.

Rehman, M., Anwar, M.A., Iqbal, M., Akhtar, K., Khalid, M.A., Ghauri, M.A., 2009. Bio-leaching of high grade Pb-Zn ore by mesophilic and moderately thermophilic iron and sulphur oxidizers. Hydrometallurgy 97, 1–7.

Saleem, A., Ebrahim, M., 2013. Production of amylase by fungi isolated from legume seeds collected in Almadinah Almunawwarah, Saudi Arabia. J. Taibah Univ. Sci. 8. https://doi.org/10.1016/j.jtusci.2013.09.002.

Segerer, A., Neuner, A., Kristjansson, J.K., Stetter, K.O., 1986. Acidianus infernus gen. nov., sp. nov., and *Acidianus brierleyi* facultatively aerobic, extremely acidophilic thermophilic sulfur-metabolizing archaebacteria. Int. J. Syst. Evol. Microbiol. 36, 559–564.

Segerer, A., Langworthy, T.A., Stetter, K.O., 1988. Thermoplasma acidophilum and Thermoplasma volcanium sp. nov. from solfatara fields. Syst. Appl. Microbiol. 10, 161–171.

Shah, A.R., Shah, R.K., Madamwar, D., 2006. Improvement of the quality of whole wheat bread by supplementation of xylanase from *Aspergillus foetidus*. Bioresour. Technol. 97, 2047–2053.

Shakoori, F.R., Aziz, I., Rehman, A., Shakoori, A.R., 2010. Isolation and characterization of Arsenic reducing bacteria from industrial effluents and their potential use in bioremediation of waste water. Pak. J. Zool. 42, 331–338.

Sharma, A., Satyanarayana, T., 2010. High maltose-forming, Ca2+-independent and acid stable α-amylase from a novel acidophilic bacterium, *Bacillus acidicola*. Biotechnol. Lett. 32 (10), 1503–1507.

Sharma, A., Parasha, D., Satyanarayana, T., 2016. Acidophilic microbes: biology and 283 applications. In: Rampelotto, P.H. (Ed.), Biotechnology of Extremophiles, Grand Challenges in Biology and Biotechnology. https://doi.org/10.1007/978-3-319-13521-2_7.

Sinha, R., Khare, S.K., 2014. Effect of organic solvents on the structure and activity of moderately halophilic *Bacillus* sp. EMB9 protease. Extremophiles. https://doi.org/10.1007/s00792-014-0683-4.

Sinsuwan, S., Rodtong, S., Yongsawatdigul, J., 2010. A NaCl-stable serine proteinase from *Virgibacillus* sp. SK33 isolated from Thai fish sauce. Food Chem. 119, 573–579.

Soni, S.K., Magdum, A., Khire, J.M., 2010. Purifification and characterization of two distinct acidic phytases with broad pH stability from *Aspergillus niger* NCIM 563. World J. Microbiol. Biotechnol. 26, 2009–2018.

Stutzenberger, F., 1987. Selective adsorption of endoglucanases from *Thermomonospora curvata* on protein-extracted lucerne fibres. Lett. Appl. Microbio. 5, 1–4.

Takayanagi, S., Kawasaki, H., Sugimori, K., Yamada, T., Sugai, A., Ito, T., Yamasato, K., Shioda, M., 1996. *Sulfolobus hakonensis* sp. nov., a novel species of acidothermophilic archaeon. Int. J. Syst. Bacteriol. 46, 377–382.

Tan, T.C., Mijts, B.N., Swaminathan, K., Patel, B.K.C., Divine, C., 2008. Crystal structure of the polyextremophilic alpha-amylase AmyB from *Halothermothrix orenii*: details of a productive enzyme-substrate complex and an N domain with a role in binding raw starch. J. Mol. Biol. 378, 852–870.

Turk, M., Mejanelle, L., Sentjurc, M., Grimalt, J.O., Gunde-Cimerman, N., Plemenitas, A., 2004. Salt-induced changes in lipid composition and membrane fluidity of halophilic yeast-like melanized fungi. Extremophiles 8, 53–61. https://doi.org/10.1007/s00792-003-0360-5.

Turk, M., Abramovic, Z., Plemenitas, A., Gunde-Cimerman, N., 2007. Salt stress and plasma-membrane fluidity in selected extremophilic yeasts and yeast-like fungi. FEMS Yeast Res. 7, 550–557. https://doi.org/10.1111/j.1567-1364.2007.00209.x.

Vaupotic, T., Plemenitas, A., 2007. Differential gene expression and HogI interaction with osmoresponsive genes in the extremely halotolerant black yeast *Hortaea werneckii*. BMC Genom. 8, 280. https://doi.org/10.1186/1471-2164-8-280.

Veana, F., Fuentes-Garibay, J.A., Aguilar, C.N., Rodriguez-Herrera, R., Guerrero-Olazaran, M., Viader-Salvado, J.M., 2014. Gene encoding a novel invertase from a xerophilic *Aspergillus niger* strain and production of the enzyme in *Pichia pastoris*. Enzym. Microb. Technol. 63, 28–33.

Vega, K., Gretty, K.V., Victor, H.S., Yvette, L., Nadia, V., Marcel, G., 2012. Production of alkaline cellulase by fungi isolated from an undisturbed rain forest of Peru. Biotechnol. Res. Int. ID 934325, 7.

Yasser, R., Abdel, F., Nadia, A., Toukhy, NM El, Gendi, H El, Rania, S.A., 2013. Production, purification and characterization of thermostable alpha amylase produced by *Bacillus licheniformis* isolate-A120. J. Chem. https://doi.org/10.1155/2013/673173.

Yim, K.J., Cha, I.T., Rhee, J.K., Song, H.S., Hyun, D.W., Lee, H.W., Kim, D., Kim, K.N., Nam, Y.D., Seo, M.J., Bae, R., Seong, W., 2014. Vulcanisaeta thermophila sp. nov., a hyperthermophilic and acidophilic crenarchaeon isolated from solfataric soil. Int. J. Syst. Evol. Microbiol. 65. https://doi.org/10.1099/ijs.0.065862-0.

CHAPTER

Screening of potential microbes for enzymes of industrial significance

3

3.1 History

Screening of enzymes or their sources is the most fundamental step in enzymology. The ultimate source of an enzyme molecule is its microbial cell. Although there are other sources such as plant and animal cells, their limitations have restricted their use. The microbial sources are generally screened from soil or water samples that behold a great diversity as seen in Chapter 2. The diversity in enzymes is directly related to its environmental niche. The desired microbes can be isolated from the selected niche favoring the attribute that is determined to pursue in the enzyme molecule. Initially, when enzymes were discovered all over the world, techniques were developed to screen microbes having the capabilities of producing a particular enzyme. The conventional method was based on using a medium in which a color change indicated the production of the enzyme of interest as the detecting agents targeted toward the enzyme was added to it. Following are the fundamental methods for isolation and screening of the enzyme-producing microbes from natural sources.

3.2 High throughput screening methods—need of present day

The industries are desirous of high potential enzymes offering resistance to the extreme environmental conditions prevailing in the process. The resistance, stability, and catalytical power are the future dimensions to be nurtured in any biocatalytic molecule.

Although nature has imparted selectivity and catalysis in every enzyme yet the sensitivity is to be handled for its proper usage. The enzyme engineering is a new horizon offering multiple options to design a molecule with additional and expected features. The tailor-made molecules are much awaited in the industrial sector as they would render trouble-free and highly economic process. The experiments on selecting the molecule with the ultimate desirous properties yield a variety of products. The screening methods play a vital role in the selection of the best combination among the available experimental outputs (Goddard and Reymond, 2004).

Protocols and Applications in Enzymology. https://doi.org/10.1016/B978-0-323-91268-6.00003-X

The screening of the desirous molecule is directed through high throughput methods of analysis and selection of the best screening of desirous molecule is presently based on the enzyme evolution pattern (Xiao et al., 2015).

The industrial market is now gradually transforming toward a biocatalyst-based process hub as it is eco-friendly as well as economical. The overall enzymes finding industrial applications are from microbial plant and animal sources. Fig. 3.1 explains the configuration of enzyme sources on a percentage basis (Sanchez and Demain, 2011; Davids et al., 2013).

Novel enzymes are therefore a constant quest of industrial fraternity. The new enzymes can be obtained by the following:

1. Enrichment culture technique (Ravuri et al., 2019)
2. Use of metagenomics (cultivating the uncultivable) (Ngara and Zhang, 2018)
3. The directed enzyme evolution (Markel et al., 2020)
4. Computational biology (in silico studies) (Araujo et al., 2012)

3.2.1 Enrichment culture

The ultimate source of microbes is nature. The enzyme, which one is attempting to find out, should be tracked in a natural environment rich in respective substrates. This approach successfully isolates microbes having natural potential for its production, which can be enhanced under laboratory conditions.

1. For example—Invertase producing microbes can be isolated from sugarcane fields, Juice Center dumping areas, molasses, and sugar factory drainage areas; xylanases producing microbes can be isolated from decaying wood, forest litter,

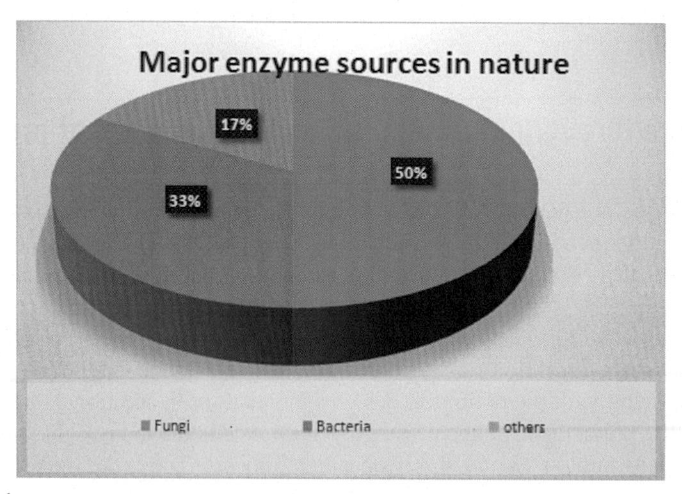

FIGURE 3.1

Major enzyme sources in nature.

or materials rich in hemicellulose and cellulosic contents. The fermentation industries are in never-ending search of production strains. The strains screened for enzyme production should have some essential features such as the enzyme yield should be high, the enzyme molecule and the microbe should be stable, and above all could be cultivable on large scale.

3.2.1.1 Preliminary screening techniques for enzyme-producing microbes

Beijernick used this technique to isolate the desired organism from soil supporting a heterogenous population. Either the physical parameters or the nutritional parameters are adjusted in such a way to favor the growth of the desired organism. For the desired enzyme-producing microbes, its substrate can be added as a sole carbon or nitrogen source whichever is applicable in context with the enzyme.

3.2.2 Uncultivables as enzyme source—a metagenomic approach

Nature harbors enormous microbes of which even less than 1% fraction is culturable (Kamagata and Tamaki, 2005). Microbes are present in nature in unimaginable number (10^{30}). The prokaryotic microbes are the most dominating among all microbial population (Turnbaugh and Gordon, 2008; Sleator et al., 2008). The genetic diversity of the prokaryotes is immense encompassing diverse and novel physiological and biochemical functions. The bioactive molecules executing such novel functions remain undetected in the laboratory. As earlier stated, the uncultivables are a plethora of microbes that escape growth in defined conditions. Many times, the samples withdrawn for microbial isolation are subjected to cultivation conditions under routine lab conditions and do not cater to the needs of most microbes that silently escape the growth and hence detection. Such unexplored genetic inventories should be explored for their richness and novelty. The incompatibility of such microbes can be attributed to exclusive growth requirements of specific pH, temperature, pressure, or even selective nutrients (Vartoukian et al., 2010).

Metagenomics has directed such genes to libraries avoiding the cultivation phase in the laboratory. Metagenomics offers two methods for screening novel genes.

3.2.2.1 On basis of sequence of the novel gene

The metagenomics is greatly dependent on the stored data it already has in the form of metagenomic libraries. The database can be very easily and rapidly accessed for comparison and analogy employing available softwares. The genetic information is screened for its probable encoded functions on grounds of data available. The output of such a method suffers immense practical drawbacks:

1. Authenticity of sequence to function relationship of the gene.
2. A single gene can cater multiple functionalities.
3. Satisfactory homology may not be found.
4. The gene may code novel product not stored in the database.

It is not possible to explore all available genetic sources for the screening of a single product. In silico studies suffer the chance of biasedness based on previous feeded data and a novel enzyme may be missed due to meager data mismatch.

3.2.2.2 On the basis of function of the novel gene

Function-based screening is more dependable if the pursuit is for a novel enzyme source. The genes are directly sourced from environmental samples for library construction. A carrier molecule is selected and introduced in the host cell for the expression of the encoded protein. The selection of the desired potent cell expressing novel gene is performed by various available throughput methods discussed elsewhere in the chapter. Fig. 3.2 provides insight into the generation of metagenomic libraries and commonly used screening methods for the desired product production. This recombinant-based functional method also offers hindrances such as incompatible vector, heterologous host, gene expression, and quantitative estimation of the target enzyme (Kim et al., 2020).

3.2.3 Directed enzyme evolution

Directed evolution is a process of enzyme rehabilitation for its efficient application in the industrial niche. The transformational process of the molecule from its natural occurrence to entirely extreme environmental conditions is a mimicry of the natural evolutionary process. The survival of the fittest is used for the selection of the best variant. The first step is a gene encoding enzyme characteristics of interest. Mutations are used

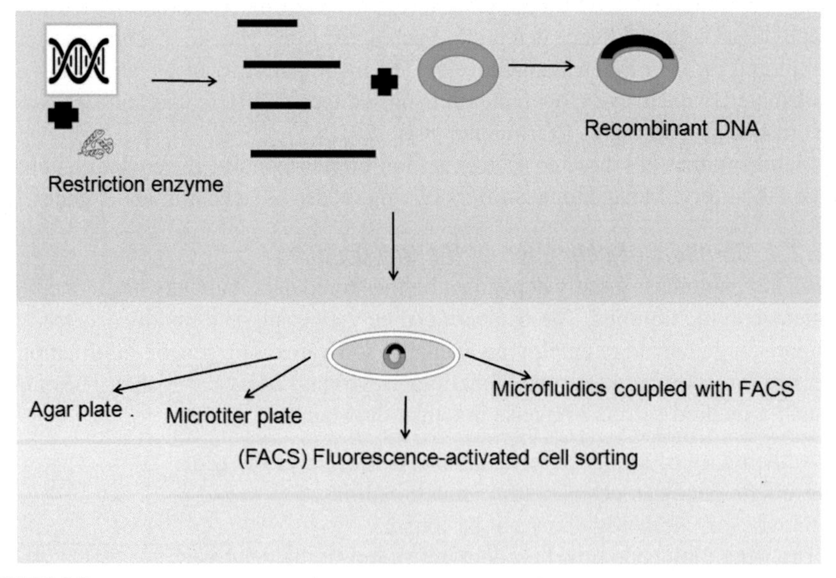

FIGURE 3.2

Metagenomic approach for exploring novel enzymes.

to elaborate a population of variants from the selected organism. These genetic variants are referred as library. The general techniques to create a library are random mutations (Lee et al., 1999), deliberate error in PCR (Abou-Nader and Benedik, 2010), saturation mutagenesis (Rocha et al., 2015), and DNA shuffling (Meyer et al., 2014). A method is employed for diverse insertions, deletions, and substitutions for unique mutants and novel enzymes. The gene library is screened for the best performing variant and is feeded as the potential cell for the next cycle of mutagenesis and selection.

This approach of modification of native enzyme characteristics for a better fit to industrial processes is referred as enzyme evolution. In this process, gene alterations are introduced randomly and the variants generated are screened for the best target.

In this process, the screening of suitable variants is crucially important. Directed evolution is explained in Fig. 3.3.

In enzyme evolution, there are two approaches—the first approach involves small changes with a desired change in the enzyme molecule. As the change is small, there is no requirement for prior information on structural and related functions.

If the desired change involves a long stretch of amino acids in the enzyme structure, then a direct approach is not feasible. There is a difficulty in generating a large size library. Therefore, the screening method should be precise and fast. The novel enzymes can be designed by the application of computational biology. In silico screening also aids to avoid the restrictions of large-size library usages (Hutter, 2018).

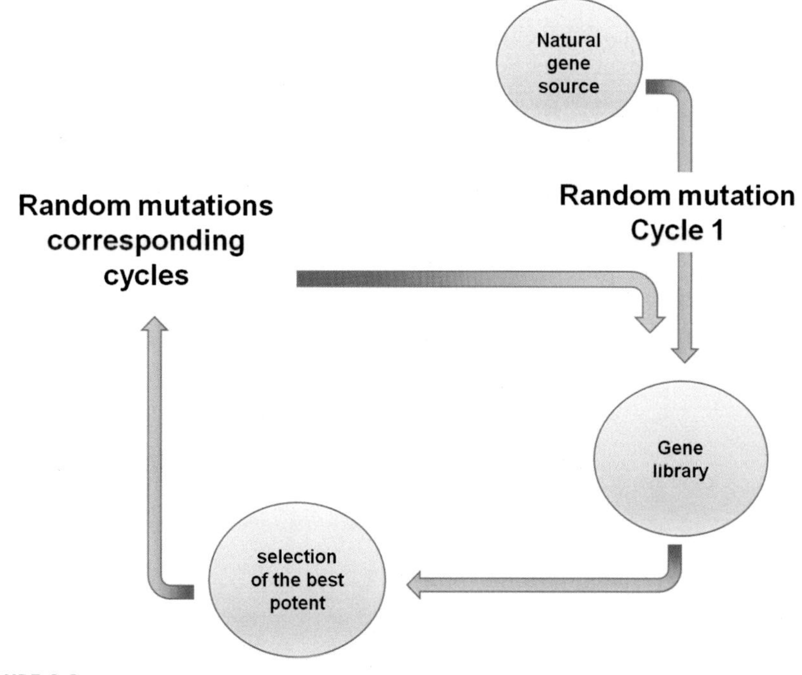

FIGURE 3.3

Directed evolution: a mimicry to natural selection (Peisajovich and Tawfik, 2007).

Although the in silico enzyme redesigning may provide alternatives for enzyme improvement, a real enzyme enhancement is observed to be purely functional. The in silico redesigned enzyme molecules sometimes fail to perform in actual reactions. Most probably, the poor performance of the redesigned enzyme is an attribute of non-consideration of active site sequences as well as its three-dimensional confirmation.

The drawback of directed evolution is that variance of all possible mutations is so high that screening of each generated variant is practically not possible. The individual screening would be time consuming and highly uneconomical in pursuit of a single enzyme.

If considered, an average enzyme polymer range is 50–600 amino acids. The scope of the library to be generated in every possible way would be unimaginable. The magnitude of this drawback (Packer and Liu, 2015; Romero and Arnold, 2009) is provided in Fig. 3.4.

These drawbacks focused the attention of scientists toward two major aspects:

1. Condensation of library size
2. Potential enhancement of throughput screening methods

Now, only those amino acids that participate in the functional site are focused, and therefore the rate of success has been improvised.

3.2.4 Computational biology (in silico studies)

The above three methods have been experimented for rediscovering enzymes with novel features. The mutational variants and their screening offer challenges yet to be met successfully.

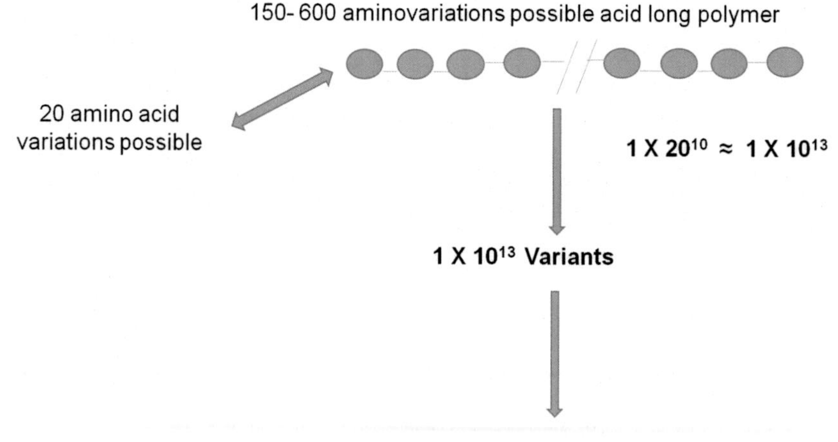

FIGURE 3.4

Drawbacks of directed evolution in enzyme screening.

Computational biology is the present discipline offering resolution to some of the problems addressed during the above-stated protocols.

Computational biology successfully offers enzyme engineering tools with the most modern computational tools.

The in silico designing mitigates the labor intensive and exhaustive lab procedures to reach about a conclusion regarding the enzyme under question (Lippow et al., 2010).

The screening of the library constructs for the desired variant of the enzyme generated the necessity of some information regarding its biochenical pathway. Computational biology has exactly bridged the gap and provides valuable assistance in designing an engineered protein. Random mutagenesis or saturation mutagenesis complemented with computational designing has improvised the field of protein engineering outputs (Treynor et al., 2007; Barderas et al., 2008).

Computational biology is a wonderful tool offering alternatives for predicting the influence of incorporation of a designed sequence of amino acids in the target proteins.

The previous approaches overlooked the mutational effect on the overall activity of the enzyme in relation to its position of occurrence. The optimum advantage can be derived by utilization of the design chemistry for improvisation of the construct and increasing the chances of targetting the best potent variant without aimless expenditure on screening all variants generated.

Such an approach was successfully used for enzyme glucose 6-oxidase. The designing was targeted to determine the influence of amino acid positioning, its interdependency, and overall impact on catalytic activity (Lippow et al., 2010).

3.3 Throughput methods for screening of enzyme variants

Ultrahigh throughput methods are defined in a generalized way as any method that can screen 10^6 compounds per day. On the contrary, the high-throughput method has a limit of 10^4 into 10^5 tests in a day (Xiao et al., 2015).

Following are the methods of screening.

3.3.1 Selection screening

This screening technique can be successfully accomplished if only the enhanced enzyme production is related to cell growth. The model cells used are *Bacillus subtilis* (Su et al., 2020), *E.coli* (Idalia and Bernardo, 2017), and *S. cerevisiae* (Gonzalez et al., 2012) for the introduction of mutant genes. Each cell expresses an individual protein variant. The cells representing distinct enzyme variant entities are screened in a selective medium. The large-sized libraries can be screened by this method. The steps involved in selection screening are presented in the flow chart (Fig. 3.5).

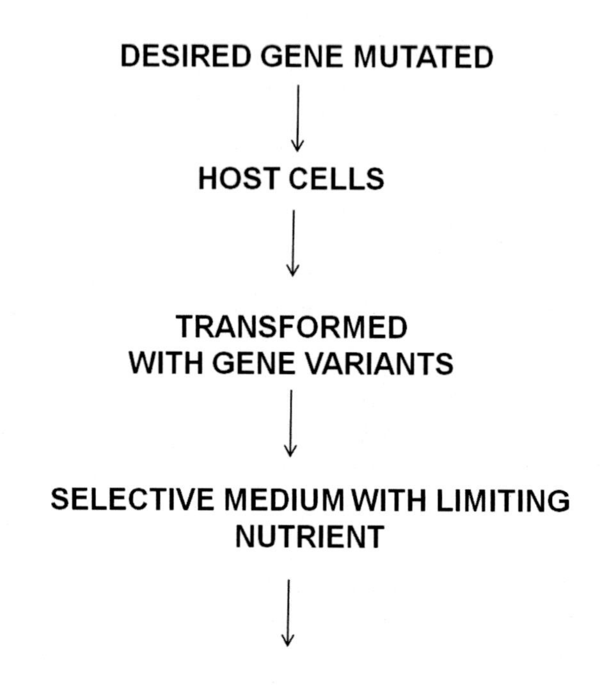

FIGURE 3.5

Overview of steps involved in selection screening.

3.3.2 Agar plate method

The agar plate method is a qualitative screening method for enzyme production by the screened cells. Quantitative estimation is not possible by this method. This procedure suffers low throughput 10^4 rounds of evolution as compared to selection screening.

This method is therefore used to screen the cells at the preliminary level whether it is a low, moderate, or potent producer. The substrate utilization or product formation produces a visible change in the medium surrounding the colony of interest evidenced by a change in color or zone of hydrolysis, etc. This method is presently used with a combination of more advanced techniques Lc-Ms/Ms to resolve environmental issues by screening remediating enzymes (Jardine et al., 2018). Fig. 3.6 exemplifies the selective screening method for enzyme amylase.

3.3.3 Microtiter plate screening method

This is one of the conventional and convenient methods allowing quantitative assay of the enzyme concerned. This method opens avenues of conjugation with a variety of analytical instruments offering an attribute of multiproduct screening (Patrick and Qian, 2013). The analytical tools successfully complemented with this technique are

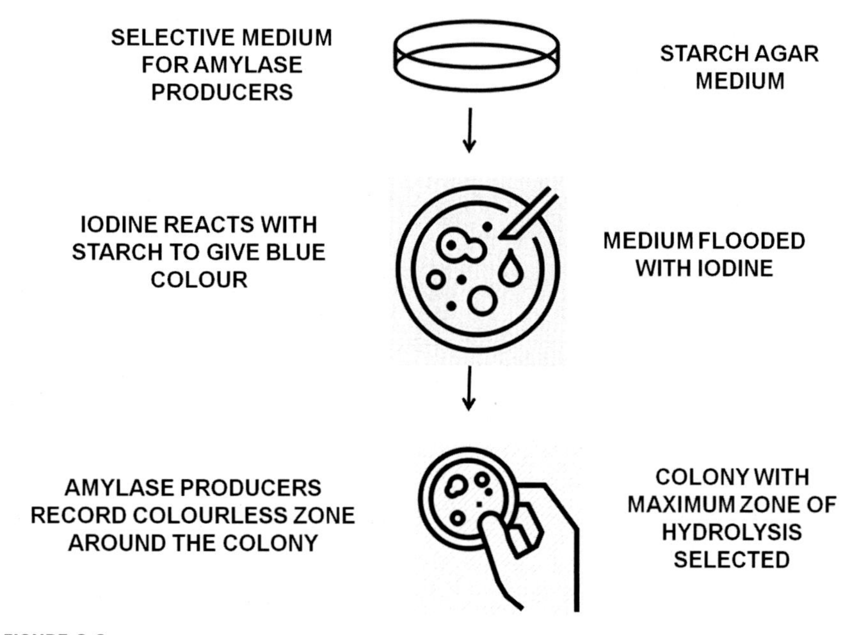

FIGURE 3.6

Selective screening method for amylase-producing organisms.

UV–visible spectrophotometry, fluorescence spectrophotometry, LCMS, and GCMS. The quantitative screening of variants provides a transparent platform for their selection. The microtitre plate (MTP) technique involves a master plate and an expression plate. The master plate is for mutant growth, and the expression plate is for cell lysis and enzyme assay.

The microtiter plates also have evidenced evolution in the design and capacity of the loaded samples. The basic principle of MTP is depicted in Fig. 3.7. Table 3.1 exemplifies the improvisation of the MTP technique by decreasing the well size and increasing the sampling tendency.

3.3.4 Fluorescent activated cell sorting

The whole concept of throughput methods is to reduce the screening reaction system, space, and increasing the number of variants screened at a single incidence. The same phenomenon has evolved the modification in microtiter plate method. The reaction system limited within the cell is scrutinized by FACS. As the screening of single-cell implies screening of a single enzyme variant, the productivity of the method is 4×10^5 cells/s.

A million or even more enzyme variants producing cells can be analyzed in <2 min. This screening procedure has enhanced the efficacy of the process from microliters to femtoliters (Zhang et al., 2019).

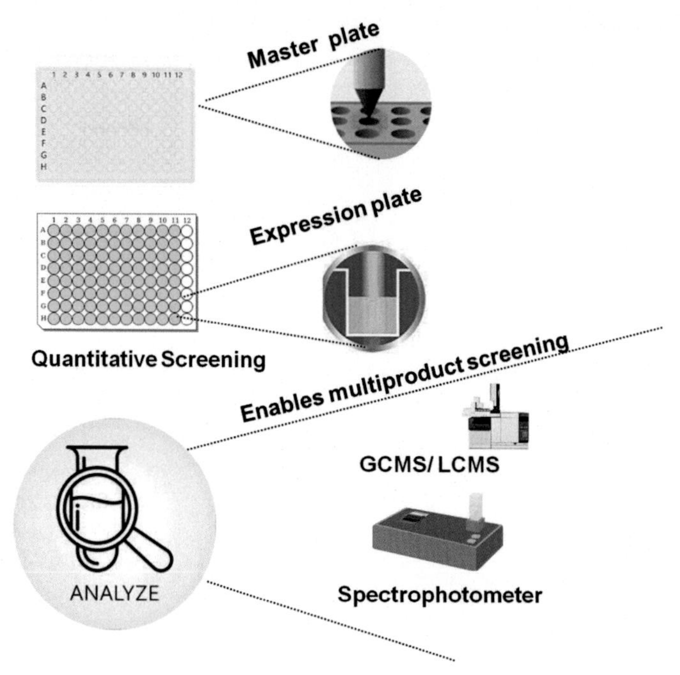

FIGURE 3.7

Advantages of microtitre plate (MTP) technique.

Table 3.1 Improvement of MTP due to miniaturization.

Wells	Bioactive entity detected	Sample size	References
96	dTDP-d-glucose4,6-dehydratase	100–200 μL	Shi et al. (2016)
384	Tyrosinase	3200 μL	Qiao et al. (2017)
1536	Low-density lipoprotein	2.5–10 μL	Knight et al. (2014)
100,000 (giga matrix)	Cells, proteins	200 nL	Robertson and Steer (2004)

The FACS is based upon screening of cells with protein variants and identifiable phenotypic traits as a marker to identify the concerned cell (Fig. 3.8).

FACs that detect enzyme activity intracellularly within the cytoplasm or extracellularly as cell surface protein display in the form of fluorescent product. The success of the use of FACS is the availability of fluorescent substrate that appropriately caters the detection necessities of FACS and the genotype–phenotype relationship.

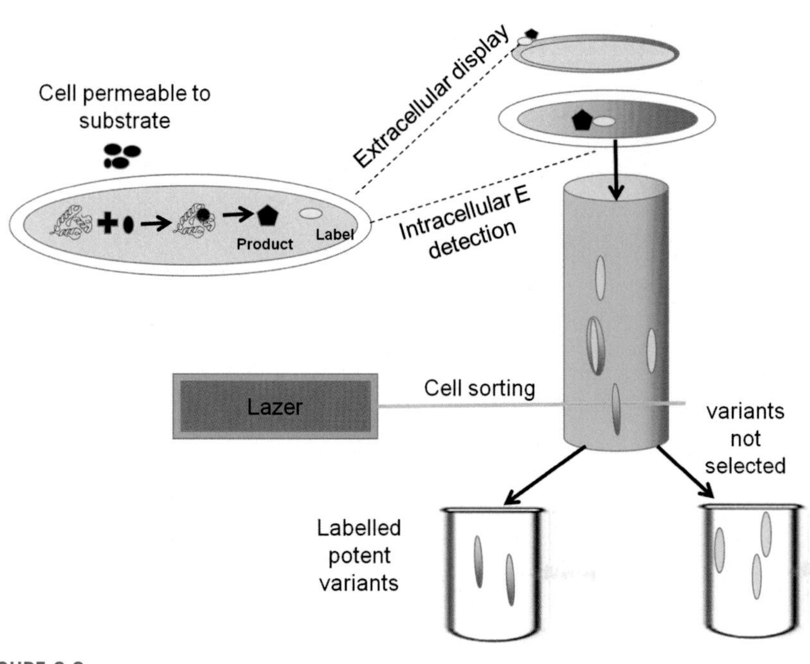

FIGURE 3.8

Fluorescence activated cell sorting (intracellular and extracellular mechanism).

Fig. 3.8 depicts extracellular protein display and intracellular display by cells and the resultant fluorescence detection and selection (cell sorting) by laser.

3.3.5 In vitro compartmentalization

This technique is also a human attempt to mimic a cell. This technique involves linking the genotype with the phenotype by artificial designing a cell-like environment with a gene and its machinery for expression. The artificial cell-like environment (referred as in vitro compartmentalization) is a surfactant mixture referred to as water-oil immersion. 1 mL of emulsion harbors nearly 10^{10} droplets or compartments catering distinct gene variant expression mixtures. The proteins expressed remain coupled with the encoding gene in the droplet that accompanies phenotype and phenotype coupling. The artificial compartmentalization has an added potential advantage of screening big libraries of width 10^8-10^{11} genes (Miller et al., 2006). The steps of preparation of artificial cell-like compartments and activity-based cell sorting are presented in Fig. 3.9.

The water-oil immersion was initially attempted for DNA by Tawfik and Griffith in 1998. This method could not be transformed to high throughput screening as the continuous oil phase is not compatible with the detection by FACS.

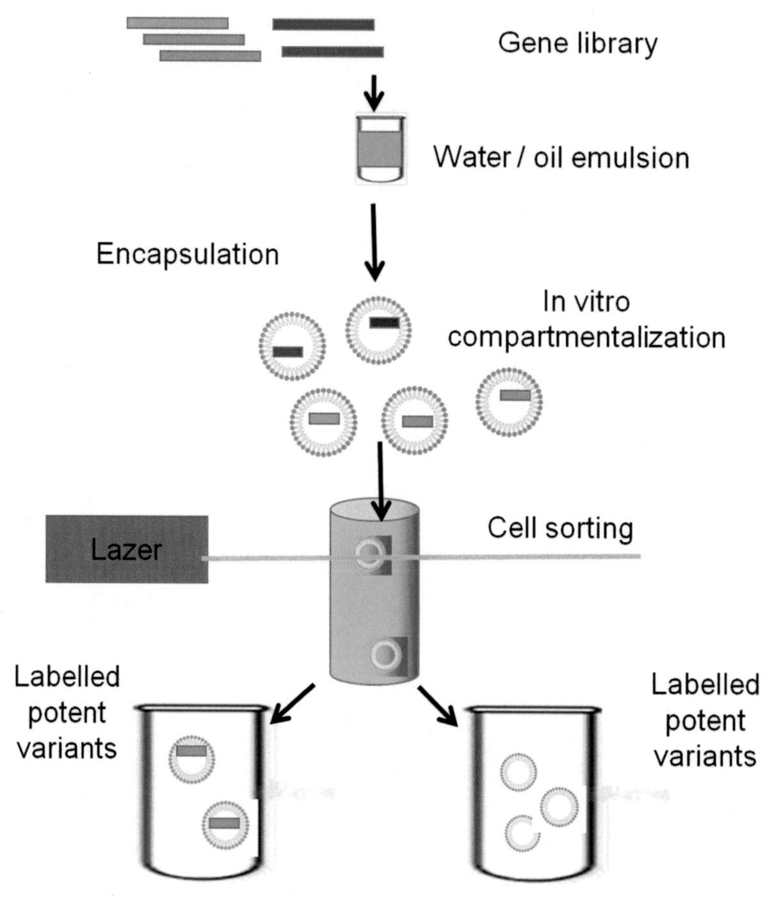

FIGURE 3.9

In vitro compartmentalization.

In vitro compartmentalization has been successfully used for enzymes beta-galactosidase (Mastrobattista et al., 2005), cellulase (Korfer et al., 2016), and hydrolases (Becker et al., 2007). To overcome the limitations of IVC, the compartmentalization materials have been widely experimented delimiting the oil emulsion usage.

3.3.6 Droplets

Microfluidics has emerged as a promising horizon in the screening of enzymes for the isolation of potent enzyme variants. This technique has successfully expanded its application through most enzymes classified under various classes. Some examples are provided in Table 3.2.

Table 3.2 Microfluidics applied to various enzyme screening.

Enzymes	References
Alpha-amylase	Sjostrom et al. (2014)
Glycoside hydrolase	Joshua et al. (2016)
(Endo-β-1,4-xylanase B), *xlnc* (endo-β-1,4-xylanase C), *cbha* (1,4-β-cellobiohydrolase A), *egla* (endoglucanase A), and *pep1* (aspartic protease), cellobiohydrolase, protease, xylanase	Beneyton et al. (2019)

The screening techniques are diverted either by detection of the enzyme (in vivo and in vitro compartments) intracellularly and extracellularly viz. by demonstrating protein on the cell surface.

The display technology exploits the extension of the gene of interest with a surface protein-encoding supplement fragment or by trapping the gene in a phage envelope. The detection process of enzyme variants is easily detectable, which attributes to the success of display methods.

3.3.7 Plasmid display

The gene of interest is supplemented with a DNA binding protein-encoding fragment. The modified enzyme with the attribute of DNA binding now conjugates with the plasmid DNA with the domain. The cell lysis releases a protein-nucleic acid complex.

3.3.8 Phage display

This technique improves the screening experiments focused on the binding efficiencies of proteins. The phage display is an efficient, fast, and popular technique used in the display for extracellular protein. The gene is fused to create a phagemid variant transferred to the appropriate host cells. The protein of interest is conjugated with enveloped protein. The packaging of phage results in the compartmentalization of genes as well as display of protein simultaneously. For expressing the target binding proteins, viruses or example M-13 are used, as they do not cause host cell lysis.

The general principle is provided in Fig. 3.10. Protein and antibody engineering has been successfully accomplished using M-13 phage (Frei and Lai, 2016). The nature of the phage selected defines whether cytoplasmic protein variants can be screened or not. This screening technique limits to evaluate the binding capabilities of enzymes but other enzyme parameters are yet to be explored by display technology. All proteins with circumvent posttranscriptional modification cannot remain explored. Retroviruses when used for eukaryotic cells enhance the protein screening efficiency (Urban et al., 2005). Various phages employed for strategic screening of proteins by phage display method are revived in Fig. 3.11.

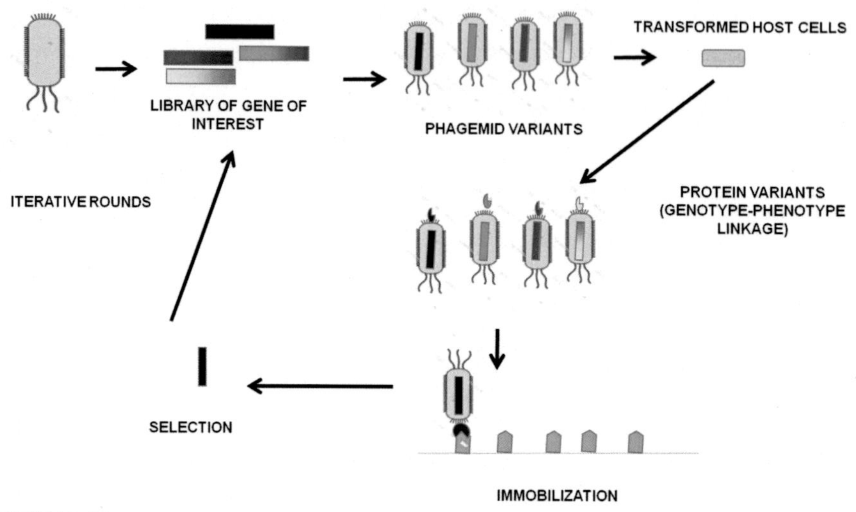

FIGURE 3.10

General principle of phage display for binding proteins using M-13 phage.

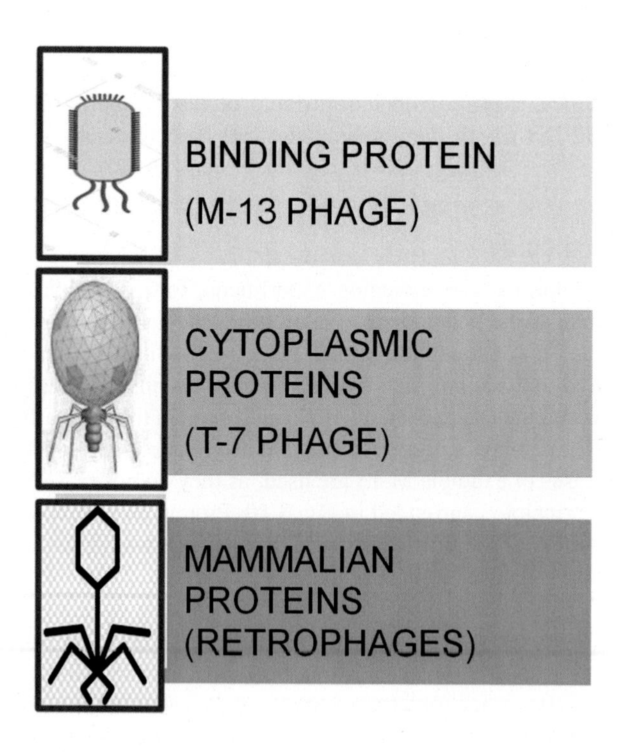

FIGURE 3.11

Phages employed for various protein displays.

3.3.9 Ribosome display m-RNA

The phenomenon of ribosome display was conceptualized by the observation of ribosome stalling (Korman et al., 1982). The initial experiment aimed at the purification of immunological proteins by simply separating the ribosomes (Hanes and Pluckthun, 1997). This technique was successfully used for immunoproteins separation with target activity. There are four main genetic components of the library design:

1. A promoter sequence.
2. An open reading frame for binding.
3. The gene of interest.
4. A spacer sequence.

In vitro transcription machinery helps in the physical synthesis of this framed library resulting in a transcript without a stop codon. In the subsequent in vitro translation, a ternary complex of m-RNA, ribosomes, and the nascent peptide attached to it are formed. The ribosome bridges the polypeptide and the m-RNA. The display protein tags from the library help in tagging and isolation during immobilization. The pictorial protocol of ribosome display is provided in Fig. 3.12.

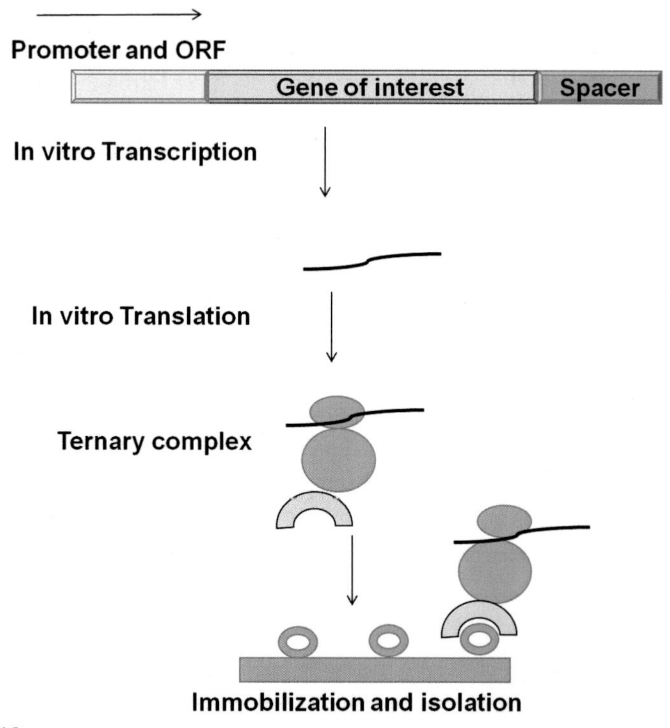

FIGURE 3.12

The mechanism of ribosome display for screening.

The screening of protein binders by this method is attempted by many (Li et al., 2019). The application has also been extended for catalytic enzymes. This method is found to be effective for both prokaryotes and eukaryotes. The success of this technique lies in the coupling of genotype and phenotype together mediated by the ribosomes.

The central theme of in vitro screening is mimicry of the cell environment. Although the transcriptional and translational machinery is made available, the prima facia question is to keep the gene and the encoded protein tagged together. In nature, the cell performs this bridging between the encoding source and final protein product rendering in vivo screening and selection easy. Artificial compartments carved with water and oil emulsion need a strategy to maintain this vital phenotype and genotype link.

The process of protein synthesis aims at the polymerization of amino acids in sequence according to the reading frame of codons encoded by the m-RNA and directed by the gene. The smaller subunit caters the codon—anticodon base pairing and affirms the positions of m-RNA and t-RNA with ribosome. The larger subunit provides passage for the nascent polypeptide outside the translational machinery.

In this process, the phenotype and genotype are linked in a sense that the m-RNA is attached to the polypeptide through t-RNA bridge at the P site of the larger subunit.

This is a temporary linkage that is dismantled at the termination of protein synthesis. The noncoding nature of termination codons and the respective release factor breaks the ester bond between t-RNA and the last amino acid. The ribosome proceeds the ribosome cycle dismantles the t-RNA and dissociates into its fraction subunits releasing the m-RNA subsequently. This junction is to be avoided if genotype-phenotype link is to be maintained in the artificial cell environment. In ribosome display technology, the absence of a stop codon avoids dissociation. The phenomenon of ribosome stalling is seen as a consequence of stop codon absentia.

3.3.10 m-RNA display

This display technology helps in the condensation of the m-RNA and protein expressed by using a linker DNA. The basic requirements remain the same: a gene library, a promoter sequence, ribosome-binding site, and open reading frame of interest devoid of a termination codon. The resultant library of mRNA molecules is introduced with DNA containing puromycin. The phenomenon of ribosome stalling is observed with a unique multimolecular complex physically linked containing the nascent polypeptide, linker DNA, puromycin, and the m-RNA. The stepwise procedure is displayed in Fig. 3.13. The generation of a new RNA ligase enzyme by using the m-RNA display method was reported by Seelig (2011).

3.3.11 The c-DNA display

The c-DNA display has alternatively been replaced by the ribosomal display that is similar to the m-RNA display that employs the covalently bonded m-RNA and protein complex. On the contrary, in the ribosomal display, the termination codon is

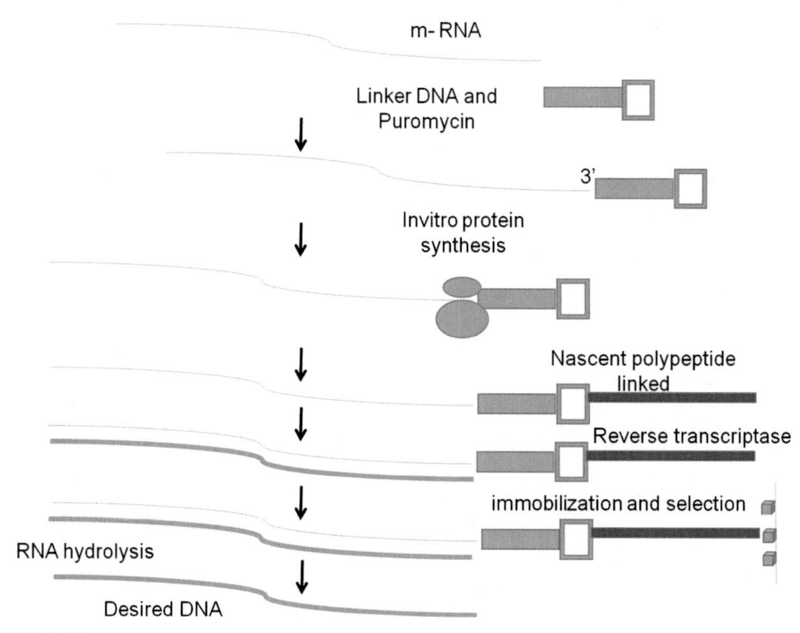

FIGURE 3.13

The mechanism of m-RNA display for screening.

eluded resulting in a sequence that bridges the protein, the unreleased t-RNA, m-RNA, and the ribosomes. In this method, puromycin plays a similar role in replacing t-RNA as in the case of m-RNA display.

All screening techniques dependent on m-RNA suffer a major drawback due to its fragile nature. The display techniques based on m-RNA are also attributed with this drawback due to exposure of the molecule to the ribonuclease enzyme.

This approach was used to eliminate the sensitivity of the m-RNA molecule toward enzyme exposure by the introduction of an additional step of C-DNA synthesis.

This improvisation leads to a stable mRNA—c-DNA—protein complex with the same principle of the use of puromycin as stated in the above method.

Although this method provides stability yet the m-RNA protein complexes should be purified faster during the subsequent steps. Successful screening of immunoglobulin G, POU-specific DNA-binding domain of Oct-1, and anti-FLAG antibody has been performed by this technique (Yamaguchi et al., 2009).

3.3.12 Reporter-based screening

When the gene activity essential for survival is used for screening the gene is termed reporter and the technique is called reporter-based screening. The critical step of this technique is to link the activity of the enzyme of interest to the reporter gene (Rossum et al., 2013).

Thus, the cells exhibiting reporter activity indirectly convey the activity of the enzyme of interest. This method screens enzymes that are not necessarily complimented to cell growth. The reporter selection method strategies are shown in Fig. 3.14.

3.3.13 Fluorescence-activated droplet sorting

This technique has been efficiently directed for the selection of enzyme variants tailored according to industrial requirements. The application of FADS for desired enzyme variant screening was described by Vallejo et al. (2019). The droplets are sorted using fluorescence sensors that detect the desired enzyme activity. The recent advances in this technique involved are discussed by Chiu and Stavrakis. (2019). A modified version of this technique has been used which is known as fluorescence lifetime-activated droplet sorting as a lab-on-a-chip devices. Successful cell sorting was performed based on the fluorescence detected in the screening droplets. This technique showed sensitivity toward substrate concentration from 1 nM to 1 mM. The sensitivity permitted segregation of droplets on basis of the mixture composition they carried within (Hasan et al., 2019). The variants of the enzyme horseradish peroxidase were successfully screened by using FADS (Agresti et al., 2010). Microfluidics for recombinant enzyme screening (Beneyton et al., 2014) is speeding up rates of enzyme discoveries (Bunzel et al., 2017); microfluidics for metagenomics study (Gielen et al., 2018); single-cell screening (Neun et al., 2019).

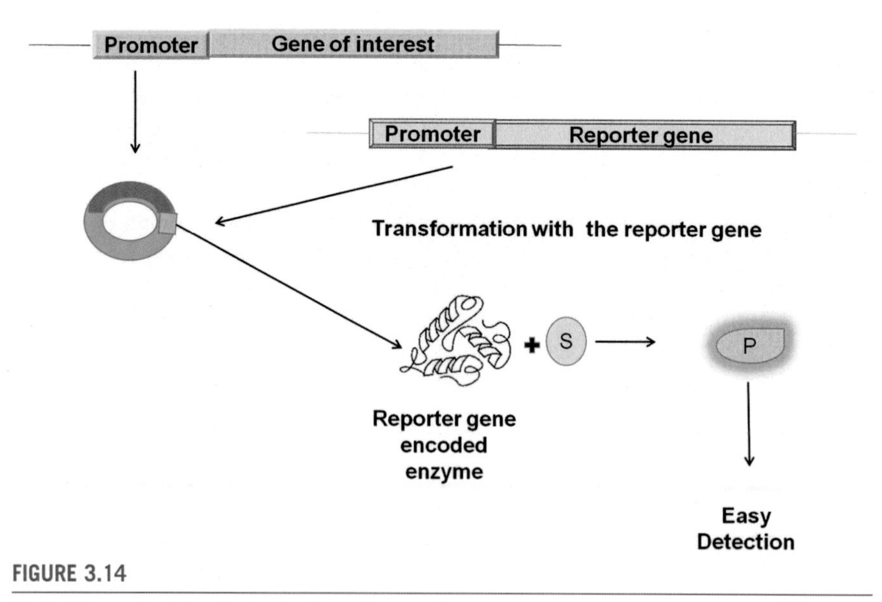

FIGURE 3.14

Underlying principle of reporter-based screening.

FADS has emerged as an interesting and promising technique of enzyme screening, as the extent of resolution is too high at the level of stereoselectivity and enantioselectivity (Lindong and James, 2019).

3.3.14 Digital imaging

Digital imaging is an alternative screening technique for an enzyme having a polymeric substrate, either low soluble substrates or result in a viscous solution rendering it unsuitable for normal throughput screening methods. Solid-phase screening is successfully used in combination with digital imaging spectroscopy for oxygenases (Joo et al., 1999). It has been successfully used for galactose oxidase Delagrave et al. (2001), transglycosidases (Kone et al., 2008), and Xiao et al. (2015). The major steps involved in this technique are given in Fig. 3.15. Pohanka (2020) has simplified the detection of enzymes by imaging the reaction by high-resolution camera in conjugation with sensing devices.

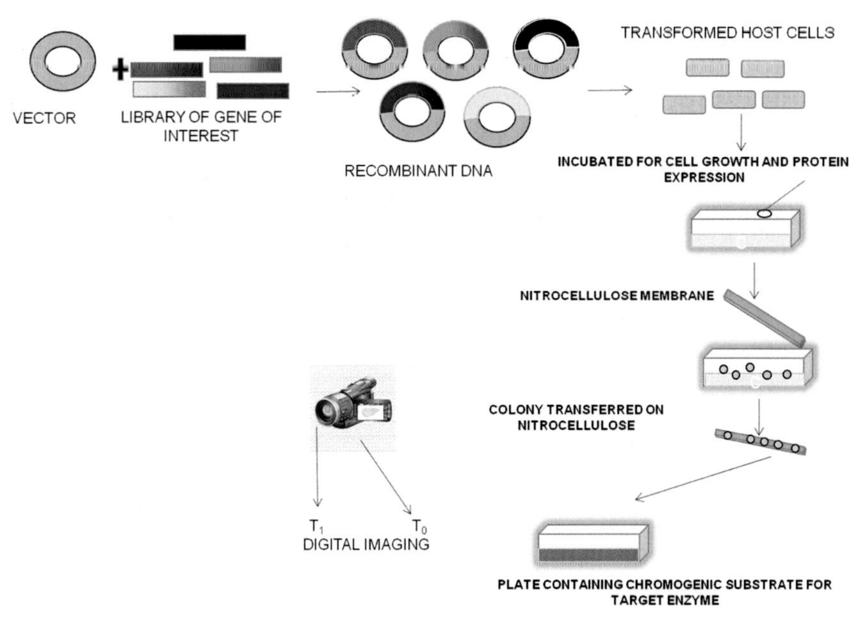

FIGURE 3.15

Screening of enzyme production by digital imaging.

3.4 Conclusion and future perspective

The future perspective of enzyme screening is promising looking toward the unexplored genetic diversity offered by the metagenomes. The need of exploring more enzymes with desirable features and industrial application edge are now pursued using computational biology as a complementing tool with the conventional screening procedures.

The present challenges lie in combatting the hindrances and incompatibility of the gene diversity with the host and its expression with absolute functionality in the selected host. The practical hurdle is the huge quantitative screening of the variants without missing the right hit.

A combination of conventional methods with modern techniques such as FACS, FADS, and microfluidics produces a clear vision during the screening and selection of the right enzyme variant.

The future is totally dependent on the high throughput methods already discussed in the chapter. The proper selection of the screening methods can successfully reduce the turmoil, misguidance, and labor behind exhaustive and seemingly endless screening protocol, ending up with a successful potent hit.

References

Abou-Nader, M., Benedik, M.J., 2010. Rapid generation of random mutant libraries. Bioeng. Bug. 1 (5), 337–340. https://doi.org/10.4161/bbug.1.5.12942.

Agresti, J.J., Antipov, E., Abate, A.R., Ahn, K., Rowat, A.C., Baret, J.C., Marquez, M., Klibanov, A.M., Griffith, A.D., Weitz, D.A., 2010. Ultrahigh-throughput screening in drop-based microfluidics for directed evolution. Proc. Natl. Acad. Sci. U.S.A. 107, 4004–4009.

Araujo, P., Sosa, B., Thomas, M., Stephen, M., 2012. In silico screening of computational enzyme designs. Protein Sci. 21 (S1), 132.

Barderas, R., Desmet, J., Timerman, P., Meloen, R., Casal, J.I., 2008. Affinity maturation of antibodies assisted by in silico modeling. Proc. Natl. Acad. Sci. U.S.A. 105, 9029–9034.

Becker, S., Michalczyk, A., Wilhelm, S., Jaeger, K.E., Kolmar, H., 2007. Ultrahigh-throughput screening to identify *E. coli* cells expressing functionally active enzymes on their surface. Chembiochem 8 (8), 943–949.

Beneyton, T., Coldren, F., Baret, J.C., Griffiths, A.D., Taly, V., 2014. CotA laccase: high-throughput manipulation and analysis of recombinant enzyme libraries expressed in *E. coli* using droplet-based microfluidics. Analyst 139 (13), 3314–3323. https://doi.org/10.1039/c4an00228h, 7.

Beneyton, T., Thomas, S., Griffiths, A.D., Nicaud, J.M., Drevelle, A., TRossignol, T., 2019. Droplet-based microfluidic high -throughput screening of heterologous enzymes secreted by the yeast *Yarrowia lipolytica*. Microb. Cell Fact 16 (1), 1–14. https://doi.org/10.1186/s12934-017-0629-5.

Bunzel, H.A., Garrabou, X., Pott, M., Hilvert, D., 2017. Speeding up enzyme discovery and engineering with ultrahigh-throughput methods. Curr. Opin. Struct. Biol. 48, 149–156. https://doi.org/10.1016/j.sbi.2017.12.010.

Chiu, F.W.Y., Stavrakis, S., 2019. High-throughput droplet-based microfluidics for directed evolution of enzymes. Electrophoresis 40 (21), 2860–2872. https://doi.org/10.1002/elps.201900222.

Davids, T., Schmidt, M., Beottcher, D., Bornscheuer, U.T., 2013. Strategies for the discovery and engineering of enzymes for biocatalysis. Curr. Opin. Chem. Biol. 17 (2), 215–220. https://doi.org/10.1016/j.cbpa.2013.02.022.

Delagrave, S., Murphy, D., Pruss, J., Maffia, A., Marrs, B., Bylina, E., Coleman, W., Grek, C., Dilworth, M., Yang, M., Youvan, D., 2001. Application of a very high-throughput digital imaging screen to evolve the enzyme galactose oxidase. Protein Eng. 14, 261–267. https://doi.org/10.1093/protein/14.4.261.

Frei, J.C., Lai, J.R., 2016. Protein and antibody engineering by phage display. Methods Enzymol. 580, 45–87. https://doi.org/10.1016/bs.mie.2016.05.005.

Gielen, F., Colin, P.Y., Mair, P., Hollfelder, F., 2018. Ultrahigh-throughput screening of single-cell lysates for directed evolution and functional metagenomics. Methods Mol. Biol. 1685, 297–309. https://doi.org/10.1007/978-1-4939-7366-8_18. PMID: 29086317.

Goddard, J.P., Reymond, J.L., 2004. Enzyme assays for high-throughput screening. Curr. Opin. Biotechnol. 15 (4), 314–322. https://doi.org/10.1016/j.copbio.2004.06.008. PMID: 15358001.

Gonzalez, P.D., Garcia, R.E., Alcalde, M., 2012. *Saccharomyces cerevisiae* in directed evolution: an efficient tool to improve enzymes. Bioeng. Bug. 3 (3), 172–177. https://doi.org/10.4161/bbug.19544.

Hanes, J., Pluckthun, A., 1997. *In vitro* selection and evolution of functional proteins by using ribosome display. Proc. Natl. Acad. Sci. U.S.A. 94 (10), 4937–4942. https://doi.org/10.1073/pnas.94.10.4937.

Hasan, S., GeisslerD., Wink, K., Hagen, A., Heiland, J.J., Belder, D., 2019. Fluorescence lifetime-activated droplet sorting in microfluidic chip systems. Lab Chip 19, 403–409.

Hutter, M.C., 2018. The current limits in virtual screening and property prediction future medicinal chemistry. Future Med. Chem. vol. 10 (13). https://doi.org/10.4155/fmc-2017-0303.

Idalia, V.M.N., Bernardo, F., 2017. *Escherichia coli* as a Model Organism and its Application in Biotechnology. https://doi.org/10.5772/67306.

Jardine, J.L., Stoychev, S., Mavumengwana, V., Ubomba-Jaswa, E., 2018. Screening of potential bioremediation enzymes from hot spring bacteria using conventional plate assays and liquid chromatography - tandem mass spectrometry (Lc-Ms/Ms). J. Environ. Manag. 223, 787–796.

Joo, H., Arisawa, A., Lin, Z., Arnold, F.H., 1999. A high-throughput digital imaging screen for the discovery and directed evolution of oxygenases. Chem. Biol. 6, 10.

Joshua, H., Kai, D., Steve, C., Shih, C., Jian, G., Paul, D., Adams, Singh, A.K., Trent, R., Northen, 2016. In-chip integration of droplet microfluidics and nanostructure-initiator mass spectrometry for enzyme screening. Lab Chip. https://doi.org/10.1039/c6lc01182a.

Kamagata, Y., Tamaki, H., 2005. Cultivation of uncultured fastidious microbes. Microb. Environ. 20 (2), 85–91.

Kim, K., Choe, D., Lee, D.H., Cho, B.K., 2020. Engineering biology to construct microbial chassis for the production of diffiffifficult-to-express proteins. Int. J. Mol. Sci. *21*, 990. https://doi.org/10.3390/ijms21030990.

Knight, S., Plant, H., McWilliams, L., 2014. Enabling 1536-Well High-Throughput Cell-Based Screening through the Application of Novel Centrifugal Plate Washing. https://doi.org/10.1002/0471142727.mb1512s105.

Kone, F.M., Le, B.M., Sine, J.P., Dion, M., Tellier, C., 2008. Digital screening methodology for the directed evolution of transglycosidases. Protein Eng. Des. Sel. 22 (1), 37–44. https://doi.org/10.1093/protein/gzn065.

Korfer, G., Pitzler, Ljubica, V., Ronny, M., Ulrich, H., 2016. In vitro flow cytometry-based screening platform for cellulase engineering. Sci. Rep. 6, 26128. https://doi.org/10.1038/srep26128.

Korman, A.J., Knudsen, P.J., Kaufman, J.F., Strominger, J.L., 1982. cDNA clones for the heavy chain of HLA-DR antigens obtained after immunopurification of polysomes by monoclonal antibody. Proc. Natl. Acad. Sci. U.S.A. 79, 1844–1848.

Lee, M.S., Dougherty, B.A., Madeo, A.C., Morrison, D.A., 1999. Construction and analysis of a library for random insertional mutagenesis in *Streptococcus pneumoniae*: use for recovery of mutants defective in genetic transformation and for identification of essential genes. Appl. Environ. Microbiol. 65 (5), 1883–1890. https://doi.org/10.1128/AEM.65.5.1883-1890.1999.

Li, R., Guangbo, K., Min, H., He, H., 2019. Ribosome display: a potent display technology used for selecting and evolving specific binders with desired properties. Mol. Biotechnol. 61. https://doi.org/10.1007/s12033-018-0133-0.

Lindong, W., James, E.S., 2019. Droplet microfluidics-enabled high-throughput screening for protein engineering. Micromachines 10, 734. https://doi.org/10.3390/mi10110734.

Lippow, S.M., Moon, T.S., Basu, S., Yoon, S.H., Li, X., Chapman, B.A., Robison, K., Lipovsek, D., Prather, K.L.J., 2010. Engineering enzyme specificity using computational design of a defined-sequence library. Chem. Biol. 17, 1306–1315.

Markel, U., Khalil, D.E., Besirlioglu, V., Schiffels, J., Streit, W.R., Schwaneberg, U., 2020. Advances in ultrahigh-throughput screening for directed enzyme evolution. Chem. Soc. Rev. *49*, 233–262.

Mastrobattista, E., Taly, V., Chanudet, E., Treacy, P., Kelly, B.T., Griffiths, A.D., 2005. High-throughput screening of enzyme libraries: in vitro evolution of a beta-galactosidase by fluorescence-activated sorting of double emulsions. Chem. Biol. 12 (12), 1291–1300.

Meyer, A.J., Ellefson, J.W., Ellington, A.D., 2014. Library generation by gene shuffling international journal of molecular sciences. Curr. Protoc. Mol. Biol. 6, 105. Unit–15.12.

Miller, O.J., Bernath, K., Agresti, J.J., Amitai, G., Kelly, B.T., Mastrobattista, E., Taly, V., Magdassi, S., Tawfik, D.S., Griffiths, A.D., 2006. Directed evolution by in vitro compartmentalization. Nat. Methods 3 (7), 561–570.

Neun, S., Kaminski, T.S., Hollfelder, F., 2019. Single-cell activity screening in microfluidic droplets. Methods Enzymol. 628, 95–112. https://doi.org/10.1016/bs.mie.2019.07.009.

Ngara, T.R., Zhang, H., 2018. Recent advances in function-based metagenomic screening. Genom. Proteom. Bioinform. 16 (6), 405–415.

Packer, M.S., Liu, D.R., 2015. Methods for the directed evolution of proteins. Nat. Rev. Genet. 16, 379.

Patrick, C.C., Shuai, Q., 2013. Protein Engineering as an Enabling Tool for Synthetic Biology Chapter 2 Huimin Zhao,Synthetic Biology. Academic Press, pp. 23–42.

Peisajovich, S.G., Tawfik, D.S., 2007. Protein engineers turned evolutionists. Nat. Meth. 4, 991–994.

Pohanka, M., 2020. Colorimetric hand-held sensors and biosensors with a small digital camera as signal recorder, a review. Rev. Anal. Chem. 3, 20–30. https://doi.org/10.1515/revac-2020-0111.

Qiao, L., Zhong, X., Belghith, E., Deng, Y., Lin, T.E., Tobolkina, E., Liu, B., Girault, G.H., 2017. Electrostatic spray ionization from 384-well microtiter plates for mass spectrometry analysis-based enzyme assay and drug metabolism screening. Anal. Chem. 89 (11), 5983–5990.

Ravuri, J.M., Maddu, S., Krishnamoorthy, G., Maram, B., Yaramolu, D., Velluri, D., 2019. Recent developments in applied microbiology and biochemistry chapter 19 - recent approaches in the production of novel enzymes from environmental samples by enrichment culture and metagenomic approach. In: Viswanath, B. (Ed.), Recent Developments in Applied Microbiology and Biochemistry. Academic Press, pp. 251–262.

Robertson, D.E., Steer, B., 2004. Recent progress in biocatalyst discovery and optimization A Steer. Curr. Opin. Chem. Biol. 8, 141–149.

Rocha, A., Reetz, M., Nov, Y., 2015. Economical analysis of saturation mutagenesis experiments. Sci. Rep. 5, 10654. https://doi.org/10.1038/srep10654.

Romero, P.A., Arnold, F.H., 2009. Exploring protein fitness landscapes by directed evolution. Nat. Rev. Mol. Cell Biol. 10, 866–876.

Rossum, T.V., Kengen, S.W.M., Oost, J.V., 2013. Reporter-based screening and selection of enzymes. FEBS 280, 2979–2996.

Sanchez, S., Demain, A.L., 2011. Enzymes and bioconversions of industrial, pharmaceutical, and biotechnological significance. Org. Process Res. Dev. 15, 224–230.

Seelig, B., 2011. mRNA display for the selection and evolution of enzymes from in vitro-translated protein libraries. Nat. Protoc. 6 (4), 540–552. https://doi.org/10.1038/nprot.2011.312. Epub 2011 Mar 31. PMID: 21455189.

Shi, X., Sha, S., Liu, L., Li, X., Ma, Y., 2016. A 96-well microtiter plate assay for high-throughput screening of *Mycobacterium tuberculosis* dTDP-d-glucose 4,6-dehydratase inhibitors. Anal. Biochem. 498, 53–58. https://doi.org/10.1016/j.ab.2016.01.004. Epub Jan 15. PMID: 26778528

Sjostrom, Bai, S.L., Yunpeng, N., Jens, J., Andersson, H.N., Svahn, H., 2014. High-throughput screening for industrial enzyme production hosts by droplet microfluidics. Lab Chip 14 (4), 806–813.

Sleator, R.D., Shortall, C., Hill, C., 2008. Metagenomics. Lett. Appl. Microbiol. 47, 361–366, 2008.

Su, Y., Liu, C., Fang, H., Zhang, D., 2020. *Bacillus subtilis*: a universal cell factory for industry, agriculture, biomaterials and medicine. Microb. Cell Fact. 19, 173. https://doi.org/10.1186/s12934-020-01436-8.

Tawfik, D., Griffith, A., 1998. Man-made cell-like compartments for molecular evolution. Nat. Biotechnol. 16, 652–656. https://doi.org/10.1038/nbt0798-652.

Treynor, T., Vizcarra, C.L., Nedelcu, D., Mayo, S.L., 2007. Computationally designed libraries of fluorescent proteins evaluated by preservation and diversity of function. Proc. Natl. Acad. Sci. U.S.A. 104, 48–53.

Turnbaugh, P.J., Gordon, J.I., 2008. An invitation to the marriage of metagenomics and metabolomics. Cell 134, 708–713.

Urban, J.H., Schneider, R.M., Compte, M., Finger, C., Cichutek, K., Álvarez-Vallina, L., Buchholz, C.J., 2005. Selection of functional human antibodies from retroviral display libraries. Nucleic Acids Res. 33 (4), e35.

Vallejo, D., Nikoomanzar, A., Paegel, B.M., Chaput, J.C., 2019. Fluorescence-activated droplet sorting for single-cell directed evolution. ACS Synth. Biol. 8 (6), 1430–1440. https://doi.org/10.1021/acssynbio.9b00103, 21.

Vartoukian, S.R., Palmer, R.M., Wade, W.G., 2010. Strategies for culture of unculturable' bacteria. FEMS Microbiol. Lett. 309, 1–7.

Xiao, H., Bao, Z., Zhao, H., 2015. High throughput screening and selection methods for directed enzyme evolution. Ind. Eng. Chem. Res. 54 (16), 4011–4020. https://doi.org/10.1021/ie503060a.

Yamaguchi, J., Naimuddin, M., Biyani, M., Sasaki, T., Machida, M., Kubo, T., Takashi, F., Yuzuru, H., Naoto, N., 2009. cDNA display: a novel screening method for functional disulfide-rich peptides by solid-phase synthesis and stabilization of mRNA–protein fusions. Nucleic Acids Res. 37, e108. https://doi.org/10.1093/nar/gkp514.

Zhang, Y., Minagawa, Y., Kizoe, H., Miyazaki, K., Iino, R., Ueno, H., Tabata, K.V., Shimane, Y., Noji, 2019. Accurate high-throughput screening based on digital protein synthesis in a massively parallel femtoliter droplet array. Sci. Adv. 5 (8), eaav8185. https://doi.org/10.1126/sciadv.aav8185.

CHAPTER

4 Solid-state fermentation and submerged fermentation for enzyme production

4.1 History of fermentation for social benefits

Enzymes are relishing exuberant industrial demand and attention due to their immediate employability as an efficient fermentation tool. Fermentations existed since the long back of human civilization. Humans then were unaware of their practice in fermentation to satisfy their specific needs. In the present scenario, the history of fermentation tracks down the lane of alcoholic beverages have been developed by humans. On the contrary, fermentation technology accompanied humans long back since 10,000 BCE. The product nurtured was fermented milk. Milk is a major dairy product of camels, goats, sheep, and cattle with self-fermentation under natural conditions. The resident microbes in milk catalyzed the first fermentation in human civilization. The environmental conditions prevailing favored the natural process and subtropical temperatures proved to be optimum for thermophilic lactic acid bacteria.

The fermentation of milk for yogurt production was carried out in pouches of goatskin loaded on camels in North Africa prevailing at high temperatures. This historic accomplishment of humans using fermentation remains unacknowledged and the process remains an unresolved mystery. However, humans initiated understanding various processes spontaneously around them in 1800.

The fermentation also catered human needs constantly without being properly apprehended till Louis Pasteur in 1856 developed a scientific linkage between the yeast and the fermentation process. He was acknowledged for establishing a scientific reason for the fermentation process rather than accepting it as a spontaneous reaction. Louis Pasteur developed a clarity of vision regarding fermentation, and worldwide, he is recognized as "Father of Fermentation Technology" The concept of fermentation was understood as a process occurring in the absence of air but parallel to all the life-supporting processes. It was not an exclusive process but an inclusive one directed by humans toward the improvement of food shelf life.

Louis Pasteur attracted the attention of the scientific community toward the microbes acting as real catalysts for the poorly understood phenomenon of

Protocols and Applications in Enzymology. https://doi.org/10.1016/B978-0-323-91268-6.00002-8

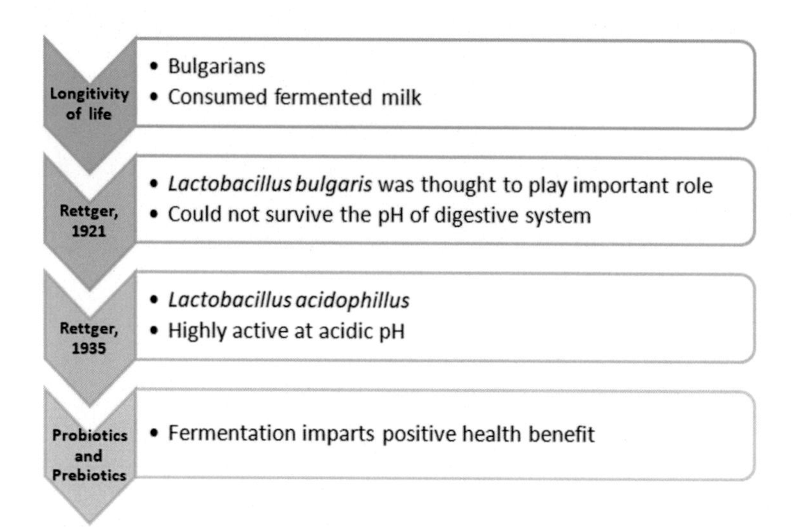

FIGURE 4.1

Beneficial effect of fermented products on human health.

fermentation. The fermentation was looked upon with limited perspectives of yielding few traditional products. The actual realm of fermentation was unleashed with the spectacular observation of Elie Metchnikoff regarding health promotion. The impact of fermented milk on the longevity of its consumers was noticed by him. The bacteria were identified and named as Bulgarian bacteria. The real research on "microbes and fermentation" initiated then. With the passage of time, humans enhanced their understanding and knowledge regarding the principle involved in fermentation. Fig. 4.1 explains the major observation of the beneficial effect of the fermented product on human health. The better understanding was expressed in the variety of fermented food products, which puffed up in society and are still prevailing with more popularity acceptance and enhanced understanding about their conferred health benefits. The concept of prebiotics, probiotics, and functional food is a virtue of all past activities that imparted the development of fermentation to present form.

4.2 Louis Pasteur and fermentation

Pasteur delivered his historic lecture in August 1857, which was later published. This lecture expressed the vision and the basis of germ theory and the role of microbes in the process. The experiment of Louis Pasteur that correlated fermentation with microbes is represented in Fig. 4.2. The controversy of cell-free conversions remained a matter of dispute among the researchers of this field.

In 1897, Buchner demonstrated ethanol production from the juice by yeast extract (in absence of living cells). He was awarded the Nobel Prize in 1907 for

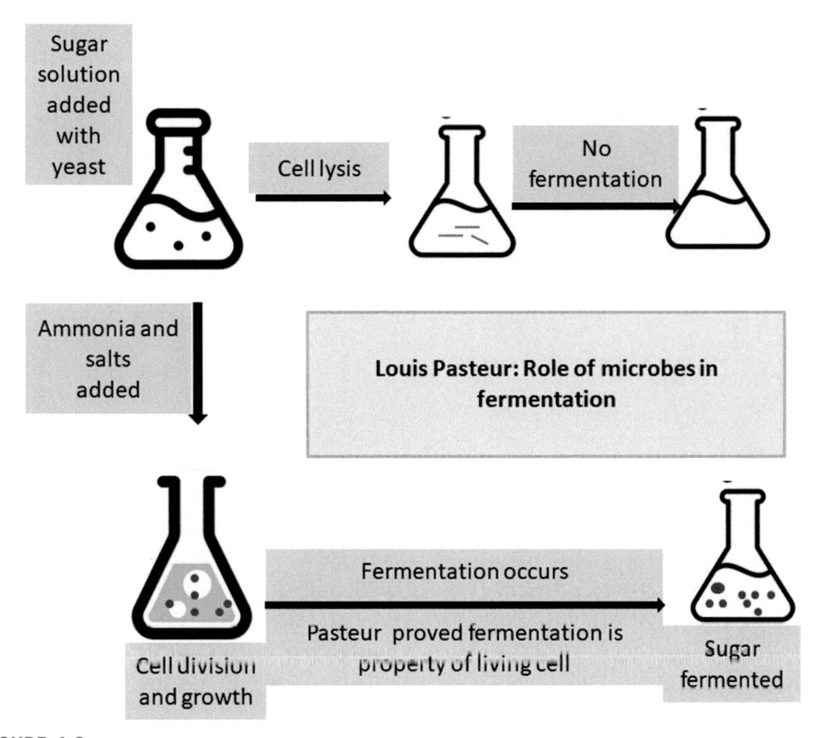

FIGURE 4.2

Role of microbes in fermentation.

his contributions (Chapter 1). Buchner's cell-free catalysis theory founded the path of the beginning of an era of enzymology at the molecular level. This historic experiment is represented in flow chart form in Fig. 4.3. It is quite evident from the previous historical events that fermentations have been consistently present since human inception.

4.3 Fermentation

The process of fermentation is a biotransformation of nutrient-rich material into an economically important product. The basic types of fermentation are indicated in Fig. 4.4. The raw materials can be secondary or by-products of dairy industry, agriculture, or various other sectors. The biotransformation is catered by microbes or enzymes produced by them.

The microbes express a variety of metabolic products during their growth curve. The typical growth curve has a lag phase, log phase, stationary phase, and decline phase. All products formed in lag and log phase are metabolites required by the cell itself for its growth and development referred as primary metabolites. The products formed in the stationary and decline phase are unnecessary by-products of the

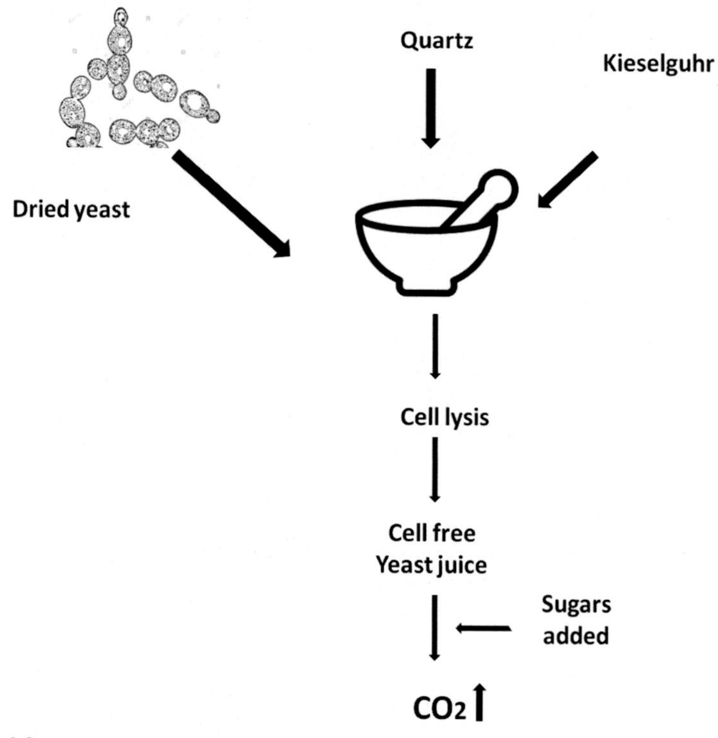

FIGURE 4.3

Cell-free fermentation: experimental proof by Buchner.

microbial metabolism and render economics virtue to the byproducts accompanying growth. Such cellular byproducts are referred as secondary metabolites. Fig. 4.5 expresses some of the primary and secondary metabolites encountered during submerged fermentation (SMF).

4.3.1 SMF fermentation

This fermentation has accompanied humans since civilization commenced. Almost all initial fermentations occurred in a liquid medium that harbored the catalyzing microbes as natural inhabitants (Ramos and Malcata, 2011). SMF is referred as liquid fermentation due to the physical state of the nutrient medium in which it is operating. Since both the constituents of the fermentation process ie., the raw materials and microbes are suspended in a liquid medium, it is referred as SMF (De Britto et al., 2020). The most common substrates naturally available are molasses, whey, corn steep liquor, and waste sulfite liquor, which are the waste products of industries related to sugar, dairy, and paper industry offering an additional benefit of waste recycling.

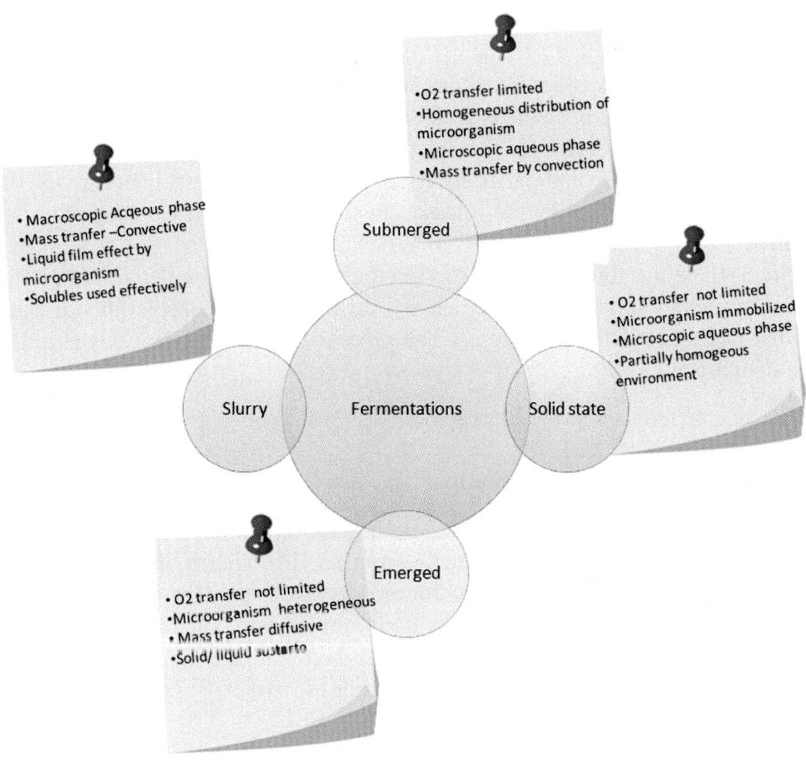

FIGURE 4.4

Types of fermentation.

Based on Glover and Kunz (2003).

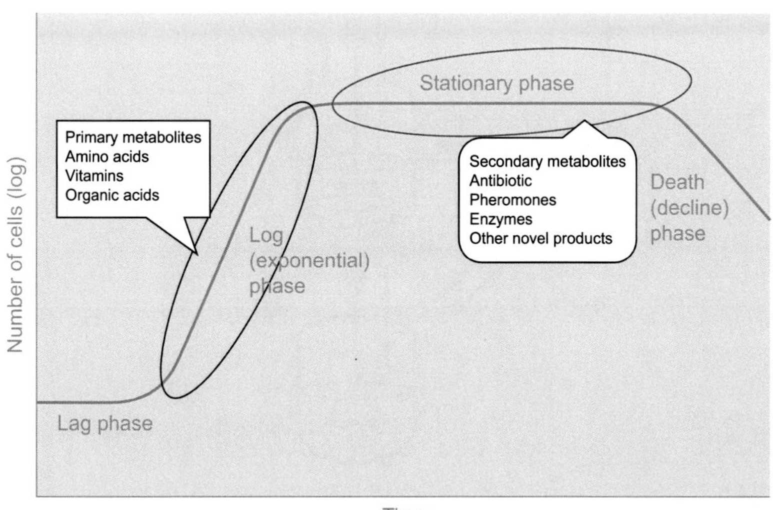

FIGURE 4.5

Growth curve and metabolite production.

As the substrates are in a fluid state, the uptake of nutrients is quite fast resulting in its exhaustion. The liquid fermentation hence requires replenishments of major energy sources in the medium after commencement of the fermentation. SMF favors an exuberant growth of bacteria as it caters the high moisture content desired for its growth. The metabolites that are secretary in nature are exuded in the liquid medium rendering their recovery and purification easier. Secondary metabolites are the most exploitable products, as they are released in the medium and can be easily down streamed.

The SMF enjoys the credit of harboring the production of a plethora of products ranging from antibiotics, anticancer agents, antimicrobial agents, antitumor agents, and many more bioactive molecules. The process of SMF can be categorized into three components:

1. The raw materials and the bioreactor
2. The microbe
3. Maintenance of optimum conditions

The selected organism is inoculated in the liquid medium and allowed to grow under optimum conditions for product formation. The nature of the raw material selected and the design of the bioreactor are determined by the growth conditions required by the organism to produce the optimum product. The fermentation conditions are vitally selected to nourish microbes as well as to maintain optimum product yield. The general layout of SMF is provided in Fig. 4.6.

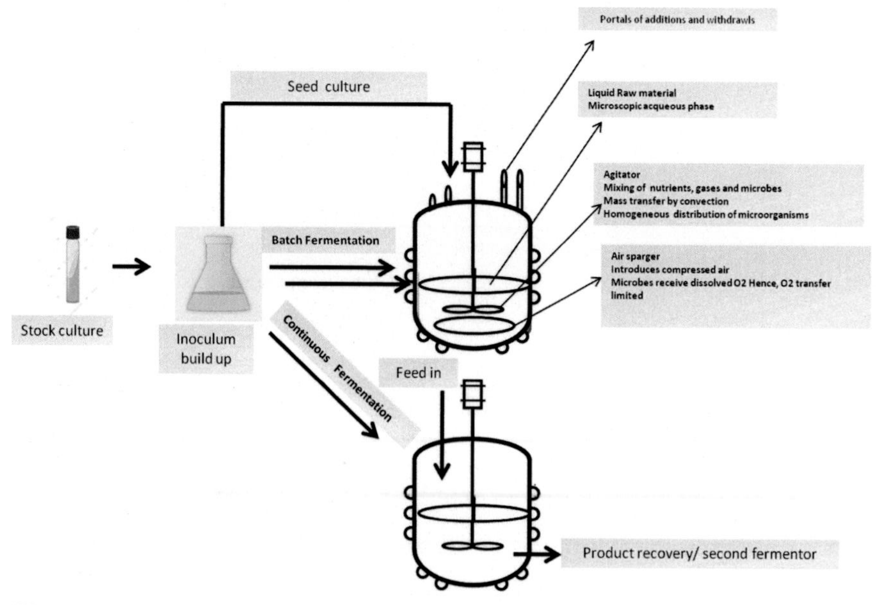

FIGURE 4.6

General layout of submerged fermentation.

The critically important factor for enzyme production using SMF is the composition of the medium or the raw material of fermentation. The raw material is the ultimate source of nutrition for the growth of selected organisms as well as product formation. The basic elemental sources that have a profound impact on growth division and enzyme synthesis are the carbon and nitrogen sources often labeled as macronutrients of the fermentation medium. The desired vitamins, inorganic salts, precursors of enzymes, and water are equally important components of the fermentation medium. Proper care must be taken to scrutinize the presence of any inhibitors in the medium that would adversely affect the productivity of the enzyme. The carbon source and nitrogen sources are of paramount importance as they initiate cell division and growth.

The raw material can be simple or complex. A simple fermentation medium is used for autotrophic microbes that consist of inorganic salt, water, and nitrogen sources. The complex medium is used for fastidious microbes that require a variety of simple and complex nutrients for their growth.

4.3.1.1 The synthetic and crude medium

A synthetic medium is a well-defined medium where its exact composition is known. The crude or complex medium like molasses and corn steep liquor contains different types of micro- and macronutrients naturally and favors the optimum growth of the microbe and hence provides optimum enzyme production. The drawbacks of SMF are the low productivity, high production cost, and complexity of the media. The advantages of SMF are as follows: the process can easily scale up, can be automated, and can minimize the influence of heat transfer limitation.

The fermentation technology and its success depend on the extent of interaction between the nutrients, gases, and microbial cells. However, the process parameters and their optimization influence the yield. The general factors often investigated are temperature, pH, incubation time, agitation, and nutritional source (Kamaran et al., 2015; Jogaiah et al., 2020). The process optimization of SMF is achieved by various approaches. Optimization often influences the yield and minimizes the processing time. Experimental designing helps to analyze the influence of independent variables and their interactive role on designing (Basri et al., 2007; Bas and Boyacı, 2007).

A more recent approach is using ANN as modeling and analyzing tool. This is a tool that has added a new dimension to modeling. It has been used in enzymology for optimization of enzyme production, study of enzyme kinetics, and study of hydrolysis of complex organic substrates as listed in Table 4.1. This tool also has become a preference in studies related to gene function, gene expression, genomics, and proteomics. Diverse data comparisons and processing are possible due to largely available processing elements (Manohar and Divakar, 2005).

The reaction rate analysis is very critical in enzymology. ANN has opened new avenues for modeling offering basic advantages like reduction in a number of experiments, common protocol for all enzyme variants, and provides solutions for complex engineering problems. Biological systems have been successfully monitored employing ANN (Bas and Boyacı, 2007). The most popular conventional method of optimization in the process is response surface methodology (RSM).

Table 4.1 The use of ANN in various fields related to enzyme production and application.

Enzymes	Employability	Ref.
Saccharifying enzymes	Bioethanol	Das et al. (2015)
Amyloglucosidase	Kinetics of enzyme	Bas et al. (2007)
Tannase	Degradation of tannin	Abdullah et al. (2020)
Hydantoinase	Medium optimization	Nagata and Chiu (2003)
Triacylglycerol acylhydrolases	Optimization of enzyme production	Parashar et al. (2019)
Thermostable lipase	Optimization of enzyme production	Ebrahimpour et al. (2008)
L-Asparginase	Optimization of enzyme production	Reddy et al. (2017)
Exochitinase	Optimization of enzyme production	Ismail et al. (2019)

This technique is statistics based proposed by Box and Wilson (1951). RSM has been used for the optimization of economically important products and has been successfully applied in microbiological and biotechnological fields (Astray et al., 2016). As a matter of fact, it is the most popular technique used in optimizing the production of microbial enzymes and has been used occasionally for maximizing the yield of microbial exochitinase (Awad et al., 2017).

4.3.1.2 General principle of submerged fermentation

This type of fermentation involves bioconversion of raw materials to products of economic importance by the involvement of microbe or enzyme derived from it. The primary and secondary metabolites are the bioactive deliverables of this process. The industrial production of bioactive molecules solely depends on the SMF to a large extent. Industrial processes render enzymes as prime fermentation products accompanied by products like vitamins and carbon dioxide gas.

The free-flowing nature of substrates generates a suspension of cells wherein the microbe absorbs nutrients and grows in the medium. Water being the major component liquefying the media, it specifically supports bacterial growth as it satisfies the high moisture demand of bacterial growth. The nutrients are in the dissolved condition in the medium. The microorganisms are homogeneously distributed in the liquid medium that forms the continuous phase. As the volume of liquid medium decides the inoculum size, 5%–10% of the media volume is generally used as inoculum size. The oxygen supply is from the dissolved oxygen hence it is limited. The mass transfer is through convection. The optimum conditions of temperature, pH can be maintained easily. SMF end with a large volume of fermented broth containing a low concentration of substrate per unit volume. The process yield is low, and recovery of the product requires high-cost input. The cost-effectiveness of the process depends on the selection of the raw material. There are two approaches to executing SMF batch-fed fermentation and continuous fermentation. Table 4.2 enlists the diverse class of enzymes produced by SMF.

Table 4.2 Microbial enzymes produced by submerged fermentation.

Microorganism	Enzyme produced	Ref.
Trichoderma harzianum	Glycosylhydrolase	Delabona et al. (2013)
Trichoderma harzianum	Glucanase	Jogaiah et al. (2013)
Trichoderma virens	Chitinase	Jogaiah et al. (2018)
Bacillus amyloliquefaciens *Bacillus* sp. *NRC22017*	Amylases	Gangadharan et al. (2008) Elmansy et al. (2018)
Aspergillus parasiticus MTCC-2796	Galactosidase	Tripathi and Mishra (2009)
Penicillium purpurogenum	α-galactosidase	Hajime et al. (1995)
Aspergillus niger JQ1516491	Glucanase	Shindia et al. (2013)
Aspergillus niger	Invertase	Romero-Gomez et al. (2000)
Aspergillus oryzae	Lactase	Shaikh (2012)
Bacillus licheniformis NCIM-2042	Proteases	Ravichandra and Shubhakar (2007)
Rhizopus chinensis	Lipase	Teng and Xu (2008)
Aspergillus niger	Pectinase	Khairnar et al. (2009) Ibrahim et al. (2015)
Aspergillus fumigatus Fresenius	Peptidase	Rodriguesda Silva et al. (2013)
Thermoascus aurantiacus	Phytases	Nampoothiri et al. (2004)
Bacillus cereus	Pullulanase	Mlook (2015)
Aspergillus sp.	Isomerase	Sayyed et al. (2010)
Debaryomyces nepalensis	Lyases	Sathyanarayana et al. (2006)
Aspergillus versicolor	Xylanases	Belorkar and Kauser (2018)
Aspergillus tubingensis CTM 507	Glucose oxidase	Mouna et al. (2020)
Cerrena unicolor C-139 *Pleurotusostreatus*	Laccases	Songulashvili et al. (2015) Mazumder et al. (2009)
Penicillium camemberti	Lipoxygenase	Husson et al. (2002)
Paenibacillus campinasensis H69-3	Glycosyltransferase	Alves-Prado et al. (2006)
Streptomyces mobaraensis	Transglutaminase	Zhang et al. (2012)
Aspergillus stallus	Fructosyltransferase	Belorkar et al. (2016)

4.3.2 Solid-state fermentation

Traditionally fermentation, especially for solid food products are commonly based on solid-state fermentation (SSF). Cheese is an excellent example of SSF. When the substrate selected for fermentation is solid, it offers a substratum for attachment and perpetuation of microbes parallel to the natural niche of microbes. Such

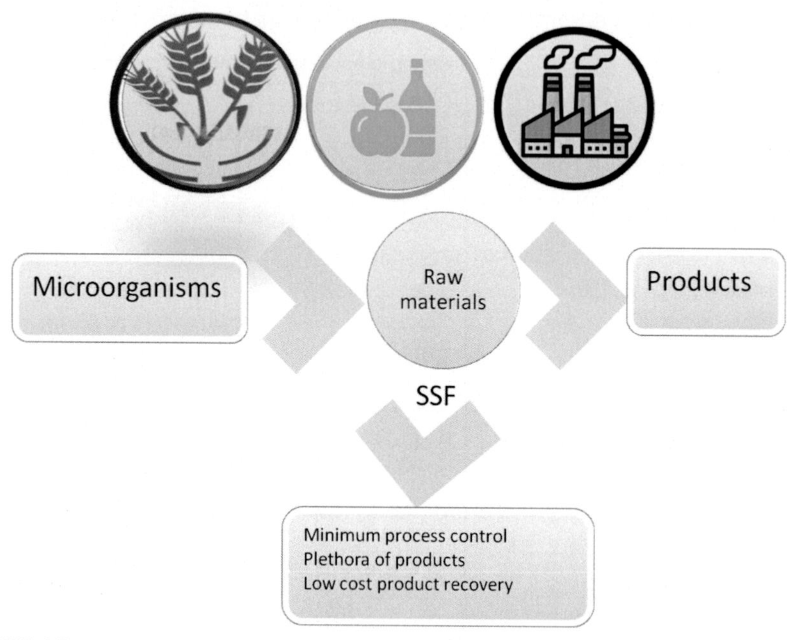

FIGURE 4.7

Major features of solid-state fermentation.

fermentations are called as SSF and they occur at extremely low water concentrations in the medium. The low moisture content of the medium favors the growth of mold and improvises its growth rate by providing a moist base for growth. The moist solid surface of the substrate proves to be an excellent growth environment for microbes.

Presently, large-scale fermentations are being successfully accomplished by using SSF. This technique has gained popularity in accord with its economic feasibility, eco-friendly nature, biomass volarization capacity, and generation of a wide range of value-added products. Fig. 4.7 reveals the major features of SSF.

Production of enzymes using SSF has now gained momentum due to accompanying economically important by-products (Satapute et al., 2019). The raw materials used are simple substrates derived as wastes of agriculture and food processing industries (Joshi et al., 2019). Apart from being economic, they provide all simple and complex nutrients essential for the support of microbial growth and enzyme production. The water-deficient environment reduces the water generated as waste to a minimum resulting in low energy demand with high productivity (Haider et al., 2018). The substrates utilized for SSF include a wide variety of complex organic materials generated as waste in agricultural fields and the industrial sector. Following Fig. 4.8 helps to understand the key features of SSF.

SSF has been considered to uphold an important place as a technique in accord with its higher productivity and stable product state for maximum enzyme

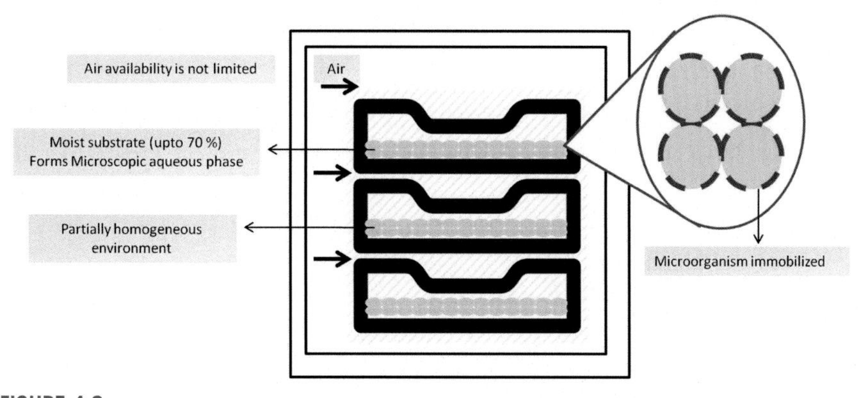

FIGURE 4.8

Basic features of solid-state fermentation.

production due to tolerance toward higher substrate and ineffectiveness of catabolite repression. The medium is complex and provides all kinds of nutrients for the growth of catalytic microbe. Aeration and process control requirements seem to be negligibly important. Water at very low concentrations offers resistance to contamination. The downstream processing cost is minimized as the product is concentrated on the surface of the solid substrate. As the absence of water eliminates the foam development henceforth antifoam agents are also not required. The main characteristics of SSF are moisture content up to 70% is within the substrate.

There is a concentration gradient due to the absorption of nutrients and production of products by the microbes immobilized on the surface of the substrate. This heterogeneity is unavoidable in SSF. The fermentation involves the interaction of gas, liquid, and solid phases among which the gas forms a continuous phase. The inoculum required is usually more than 10% in contrary to solid SMF. The inoculums should reach to each and every particle of the substrate. As microbes are easily accessible to the gaseous oxygen trapped in between the substrates it results in cost reduction otherwise required for oxygen availability. SSF provides a promising option for potential microbes that portray sensitivity toward the shear stress and incapable of withstanding agitation. It is difficult to monitor and control the process parameters, and the growth of microbes is solely dependent on nutrient diffusion and accessibility. The basic outlay of SSF is provided in Fig. 4.8.

4.3.2.1 Bioreactors used in solid-state fermentation

An insight for the essential features that a bioreactor should cater if it is intended for SSF (Mitchell et al., 2019). The bioreactor design was detailed and three essential features were proposed for consideration while designing any specific bioreactor. Any bioreactor design should be able to handle the heat generation during the process, balance the water content at any point during fermentation progress and should provide satisfactory oxygen to all cells catalyzing the bioconversion.

The main criteria for the selection of bioreactor are as follows:

1. Method used for agitation.
2. Method used for aeration.

On the basis of the above two parameters, four classes of bioreactors are employed in SSF.

a. Tray-type bioreactors
b. Packed-bed bioreactors
c. Rotating- or stirred-drum bioreactors
d. Fluidized bed type

4.3.2.2 Tray-type bioreactors

These bioreactors comprise rectangular trays inside which thin layers of substrate are maintained. This bioreactor is simple, traditional, and provides a uniform, static bed of substrate for the growth of microbes. The size of the substrate is crucially important as it will decide the pore size entrapped (medium size) or the compaction generated (too small). This bioreactor follows the static method of fermentation. As this bioreactor does not involve mechanical agitation, the trays are perforated at the bottom to minimize the effect of heat generated in the bed. Multiple trays can be placed one above the other with considerable space in between. The trays are kept in an incubation chamber to provide the optimum temperature. Humidity controls can also be added to the incubation chamber. Air is allowed to flow freely but not forced into it. Table 4.3 creates an overview of the type of enzyme produced by the tray type of bioreactor.

Table 4.3 Enzymes produced by solid-state fermentation using tray-type bioreactor.

Enzyme produced	Microbe involved	Inert support used in tray type
Pectinase	*Aspergillus niger Aa-20*	Lemon peel pomace (Ruiz et al., 2012)
Lipase	*Rhizopus oryzae* *Penicilliums implicissimum*	Sugarcane baggasse (Vaseghi et al., 2012) Babassu cake (Gutarra et al., 2005)
L-asparaginase	Aspergillus sp.	Cottonseed cake, wheat bran, and red gram husk (Doriya and Kumar, 2018)
Alkaline protease	*Aspergillus oryzae PTCC 5164*	Rice bran, wheat bran, soybean meal, and wheat flour (Maryam and Mohammad, 2015)
cellulolytic enzymes	*Trichoderma reesei* and *Aspergillus oryzae*	Soybeans hulls (Brijwani, 2011)

4.3.2.3 Packed beds bioreactors

Packed-beds bioreactors are column reactors. The material used to build the reactor is glass or plastic. The column is packed with the selected substrate on the column base that has perforations. The base is used for introducing air under pressure inside the column of the substrate. This design facilitates efficient process control and heat removal. Alternative approaches of moist air circulations efficiently handle the temperature rise. These bioreactors have intermediate stirring that minimizes the heterogeneous environment of SSF. The growth of microbe is heterogeneous. The temperature of water circulated for cooling is to be monitored for balancing the temperature changes due to fermentation. As the bed is packed in the column cake formation in the bed is possible. The major drawbacks are difficulties in scale-up and downstream processing of the product. Packed bed type of bioreactors has been employed for enzyme production using SSF (Table 4.4).

4.3.2.4 Rotating drum bioreactors

Rotating drum bioreactors are also referred as stirred drum bioreactors. The mixing is at intervals with no forced aeration. The bioreactor is drum or cylinder placed horizontally and filled with a shallow layer of the substrate. The use of minimized substrate renders more headspace free for the occupancy of air. This allows a free spatial interaction between air, the substrate, and the microbe growing on it. The drum is rotated slowly to accomplish desired mixing and aeration yet not disturbing the microbial growth on the substrate surface. The process variation is used to

Table 4.4 Enzymes produced by solid state fermentation using packed bed-type bioreactor.

Enzyme produced	Microbe involved	Inert support used in packed bed
Tannase	*Aspergillus niger*	Polyurethane foam (Rodríguez-durán et al., 2011)
Glucoamylase	*Aspergillus niger*	Substrate packed (Pandey and Radhakrishnan, 1992)
	Penicillium javanicum	Solanum tuberosum L. waste (El-Gendy and Alzahrani, 2020)
Pectinases Lipases	*A. niger* and *A. oryzae*	Wheat bran and sugarcane bagasse (Mitchell et al., 2019)
Lipase	*Penicilliums implicissimum*	Babassu cake with cane molasses (Cavalcanti et al., 2005)
Cellulase	*Myceliophthora thermophile*	Sugarcane bagasse (Perez et al., 2019)
B-Mannanase	*Aspergillus niger F12.*	Coffee industry waste (Favaro et al., 2020)
Endoglucanase xylanase	*Myceliophthora thermophila I-1D3b*	Sugarcane bagasse and wheat bran (Perez et al., 2020)

Table 4.5 Enzymes produced by solid-state fermentation using rotating drum-type bioreactor.

Enzyme produced	Microbe involved	Inert support used in rotating drum
Ligninolytic enzyme	*Phanerochaete chrysosporium*	Cubes of nylon sponge (Domınguez et al., 2001)
Fucoidanase	*Aspergillus niger PSH* and Mucor sp. 3P	*Fucus vesiculosus* algal biomass (Rodríguez-Jasso et al., 2013)
Pectinases Xylanases Cellulases	*Aspergillus awamori*	White grape pomace (Díaz et al., 2009)
Pectinase	*A. niger*	Orange pomace (Mahmoodi et al., 2019)
Ligninolytic enzyme	*Phanerochaete chrysosporium*	Inert support (Dominguez et al., 2001)
Cellulases Xylanase	*Aspergillus niger*	Sugarcane bagasse, Palm kernel (Lee et al., 2017)

optimize the bioconversion inside the rotating drum bioreactor. The mixing employed can be continuous, at short intervals or after long intervals. Forced aeration can be used to balance the temperature fluctuations. Air used can be conditioned to minimize temperature heterogeneity. The heat generated in the bed diffuses through the bed and moves toward the headspace. There are advantages of intermittent mixing as it allows efficient mass transfer as well as conserves intact mycelia growth. There are varied examples of fermentations producing commercially important enzymes using rotating drum type bioreactors (Table 4.5).

4.3.2.5 Fluidized-bed bioreactors

This type of bioreactor has a vertical cylinder as the main body of the fermentation vessel. The base of bioreactors has pores. The air is forced from the base that provides liquidity to the bed. This fluidity causes efficient mixing of all three components: solids, liquids, and gases. The liquid content maintains the temperature of the column imparting a cooling effect due to water evaporation. Thus, the heating effect is neutralized by the evaporation of water in the liquid phase.

The particle size of the solids selected is critically important as it determines the buoyancy and fluidity of the bed. Large-sized substrate particles are undesirable as they may have a tendency to flocculate and settle down decreasing the fluidity of the bed. The fluidized bed reactors are excellent examples of efficient heat and mass transfer. This bioreactor design offers a major restriction in form of the height/radius ratio of the column, lag time required for biofilm development. Recycling causes a major energy deficit increasing the cost of operation.

Fluidized bed reactors help in the execution of various enzymes applications. The major thrust area for its scope is environmental engineering. As the process operates on the hydrodynamic principle, it presents the advantage of high efficiency. Recycling totally eludes the microbes from the metabolite/repressors inhibition effect and conserves it for future cycles of bioconversions (Ozkaya et al., 2019).

4.4 Conclusion and future directions

SMFs are presently widely used for the commercial production of enzymes. The supreme role in SMF success on the industrial front is of bioreactors. The industries stress upon the accurate selection of incumbent components, that is, the bioreactor, raw material, and potential microbe of SMF for maximizing yield and minimizing production cost. Future research must be directed toward the management of heat transfer, mass transfer, and a negligible level of contamination in the process.

Patents of processes catering have been initiated since past 2 decades and are filed continued till date. The future research is engrilled in the present patents directed more toward the better remodeling of bioreactors rather than improvising process parameters. SSF is to be targeted more for its economic feasibility and efficient product formation. Research should be focused to reduce heterogeneity and improve process parameter control.

References

Abdullah, S., Pradhan, R.C., Aflah, M., Mishra, S., 2020. Efficiency of tannase enzyme for degradation of tannin from cashew apple juice: modeling and optimization of process using artificial neural network and response surface methodology. J. Food Process. Eng. https://doi.org/10.1111/jfpe.13499.

Alves-Prado, H.F., Gomes, E., Silva, R.D., 2006. Evaluation of solid and submerged fermentations for the production of cyclodextrin glycosyltransferase by *Paenibacillus campinasensis* H69-3 and characterization of crude enzyme. Appl. Biochem. Biotechnol. 234–246.

Astray, G., Gullón, B., Labidi, J., Gullón, P., 2016. Comparison between developed models using response surface methodology (RSM) and artificial neural networks (ANNs) with the purpose to optimize oligosaccharide mixtures production from sugar beet pulp. Ind. Crop. Prod. 92, 290–299.

Awad, G.E., Wahab, W.A.A., Hussein, M., El-diwany, A., Esawy, M.A., 2017. Sequential optimizations of Aspergillus awamori EM66 exochitinase and its application as biopesticide. J. Appl. Pharmaceut. Sci. 7, 67–75.

Bas, Boyacı, İ.H., 2007. Modeling and optimization II: comparison of estimation capabilities of response surface methodology with artificial neural networks in a biochemical reaction. J. Food Eng. 78, 846–854. https://doi.org/10.1016/j.jfoodeng.2005.11.025.

Bas, D., Dudak, F.C., Boyac, I.H., 2007. Modeling and optimization III: reaction rate estimation using artificial neural network (ANN) without a kinetic model. J. Food Eng. 79, 622–628.

Basri, M., Rahman, R.N.Z.R.A., Ebrahimpour, A., Salleh, A.B., Gunawan, E.R., Rahman, M.B.A., 2007. Comparison of estimation capabilities of response surface methodology (RSM) with artificial neural network (ANN) in lipase-catalyzed synthesis of palm-based wax ester. BMC Biotechnol. 7, 1–14. https://doi.org/10.1186/1472-6750-7-53.

Belorkar, S.A., Gupta, A.K., Rai, R., 2016. Enhancement of extracellular fructosyltransferase production by Aspergillus stallus through batch fermentation. J. Pure. Appl. Microbiol. 10, 649–655.

Belorkar, S.A., Kausar, H., 2018. Prelimnary screening of molds for production of xylanase. J. Pure Appl. Microbiol. 12, 161–164. https://doi.org/10.22207/JPAM.12.1.20.

Box, G.E., Wilson, K.B., 1951. On the experimental attainment of optimum conditions. J. Roy. Stat. Soc. B 13 (1), 1–38.

Brijwanik, 2011. Solid State Fermentation of Soybean Hulls for Cellulolytic Enzymes Production: Physicochemical Characteristics, and Bioreactor Design and Modeling. University of Delhi, India, 2000 Department of Grain Science & Industry College of Agriculture Kansas State University, Manhattan, Kansas. PhD Thesis.

Cavalcanti, E.D.C., Gutarra, M.L.E., Freire, D.M.G., Castilho, L.D.R., Júnior, G.L.S., 2005. Lipase production by solid-state fermentation in fixed-bed bioreactors. Braz. Arch. Biol. Technol. 48, 79–84. https://doi.org/10.1590/S1516-89132005000400010.

Das, S., Bhattacharya, A., Haldar, S., Ganguly, A., Gu, S., Ting, Y.P., Chatterjee, P.K., 2015. Optimization of enzymatic saccharification of water hyacinth biomass for bio-ethanol: comparison between artificial neural network and response surface methodology. Sust. Mat. Technol. 3, 17–28.

De Britto, S., Tanzeembanu, D.G., Praveen, S., Lalitha, S., Ramachandra, Y.L., Jogaiah, S., Ito, S., 2020. Isolation and characterization of nutrient dependent pyocyanin from *Pseudomonas aeruginosa* and its dye and agrochemical properties. Sci. Rep. 10, 1542. https://doi.org/10.1038/s41598-020-58335-6.

Delabona, P.S., Farinas, C.S., Lima, D.J.S., Pradella, J.G.C., 2013. Experimental mixture design as a tool to enhance glycosyl hydrolases production by a new *Trichoderma harzianum* P49P11 strain cultivated under controlled bioreactor submerged fermentation. Bioresour. Technol. 132, 401–405.

Díaz, A., Ignacio, O., Ildefonso, C., Ana, B., 2009. Solid state fermentation in a rotating drum bioreactor for the production of hydrolytic enzymes. Chem. Eng. Transact. 17. https://doi.org/10.3303/CET0917174.

Domınguez, A., Rivela, I., Couto, S.R., Sanromán, M.A., 2001. Design of a new rotating drum bioreactor for ligninolytic enzyme production by *Phanerochaete chrysosporium* grown on an inert support. Process Biochem. 37, 549–554.

Doriya, K., Kumar, D.S., 2018. Optimization of solid substrate mixture and process parameters for the production of L-asparaginase and scale-up using tray bioreactor. In: Biocatalysis and Agricultural Biotechnology, vol. 13, pp. 244–250.

Ebrahimpour, A., Rahman, R.N.Z.R.A., Ean Ch'ng, D.H., Basri, M., Abu, B.S., 2008. A modeling study by response surface methodology and artificial neural network on culture parameters optimization for thermostable lipase production from a newly isolated thermophilic *Geobacillus* sp. strain ARM. BMC Biotechnol. 8, 96. https://doi.org/10.1186/1472-6750-8-96.

El-Gendy, M.M.A.A., Alzahrani, N.H., 2020. Solid state fermentation of agro-industrial residues for glucoamylase production from endophytic fungi *Penicillium javanicum* of

Solanum tuberosum L. J. Microb. Biochem. Technol. 12, 426. https://doi.org/10.35248/1948-5948.20.12.426.

Elmansy, E.A., Asker, M.S., El-Kady, E.M., et al., 2018. Production and optimization of α-amylase from thermo-halophilic bacteria isolated from different local marine environments. Bull. Natl. Res. Cent. 42, 31. https://doi.org/10.1186/s42269-018-0033-2.

Favaro, C.P., Baraldi, I.J., Casciatori, F.P., 2020. β -Mannanase Production using coffee industry waste for application in soluble coffee processing. Biomolecules 10 (2), 227. https://doi.org/10.3390/biom10020227.

Gangadharan, D., Sivaramakrishnan, S., Nampoothiri, K.M., Sukumaran, R.K., Pandey, A., 2008. Response surface methodology for the optimization of alpha amylase production by Bacillus amyloliquefaciens. Bioresour. Technol. 99, 4597–4602.

Glover, A.K., Kunz, B., 2003. Theoritical characterization of slurry fermentation. Eng. Life Sci. 3 (5), 237–241.

Gutarra, M.L., Cavalcanti, E.D., Castilho, L.R., Freire, D.M., Sant'Anna, G.L., 2005. Lipase production by solid-state fermentation: cultivation conditions and operation of tray and packed-bed bioreactors. Appl. Biochem. Biotechnol. 121–124, 105–116. Springer.

Haider, M.S., Kurjogi, M., Fiaz, M., Jogaiah, S., Wang, C., Fang, J., 2018. Drought stress revealed physiological, biochemical and gene-expressional variations in 'Yoshihime' peach (*Prunus Persica* L) cultivar. J. Plant Interact. 13, 83–90.

Hajime, S., Hideyuki, K., Gwi, G., Yoko, K., Taku, S., Reiji, K., Hiroaki, N., Shigeki, Y., Kunihiro, K., Isao, K., 1995. Purification and some properties of α-galactosidase from *Penicillium purpurogenum.* Bioscience. Biotechnol. Biochem. 59, 2333–2335.

Husson, F., Thomas, M., Kermasha, S., Belin, J.M., 2002. Effect of linoleic acid induction on the production of 1-octen-3-ol by the lipoxygenase and hydroperoxide lyase activities of *Penicillium camemberti.* J. Mol. Catal. B Enzym. 19, 363–369.

Ibrahim, D., Weloosamy, H., Lim, S.H., 2015. Effect of agitation speed on the morphology of *Aspergillus niger* HFD5A-1 hyphae and its pectinase production in submerged fermentation. World J. Biol. Chem. 6 (3), 265–271. https://doi.org/10.4331/wjbc.v6.i3.265.

Ismail, S., Serwa, A., Abood, A., Fayed, B., Ismail, S., Hashem, A., 2019. A study of the use of deep artificial neural network in the optimization of the production of antifungal exochitinase compared with the response surface methodology. Jord. J. Appl. Sci. 12, 541–551.

Jogaiah, S., Mostafa, A., Tran, L.S.P., Shin-ichi, I., 2013. Characterization of rhizosphere fungi that mediate resistance in tomato against bacterial wilt disease. J. Exp. Bot. 64, 3829–3842.

Jogaiah, S., Abdelrahman, M., Tran, L.-S.P., Ito, S.-I., 2018. Different mechanisms of *Trichoderma virens*-mediated resistance in tomato against Fusarium wilt involve the jasmonic and salicylic acid pathways. Mol. Plant Pathol. 19, 870–882.

Jogaiah, S., Praveen, S., De Britto, S., Konappa, N., Udayashankar, A.C., 2020. Exogenous priming of chitosan induces upregulation of phytohormones and resistance against cucumber powdery mildew disease is correlated with localized biosynthesis of defense enzymes. Int. J. Biol. Macromol. 162, 1825–1838.

Joshi, S.M., De Britto, S., Jogaiah, S., Ito, S., 2019. Mycogenic selenium nanoparticles as potential new generation broad spectrum antifungal molecules. Biomolecules 9 (9), 419.

Kamaran, A., Bibi, Z., Aman, A., Qader, S.A.U., 2015. Lactose hydrolysis approach: isolation and production of b- Galactosidase from newly isolated Bacillus strain B-2. In: Biocatalysis and Agricultural Biotechnology, vol. 12, pp. 1–19.

Khairnar, Y., Krishna, K.V., Boraste, A., Gupta, N., Trivedi, S., Patil, P., Gupta, G., Gupta, M., Jhadav, A., Mujapara, A., Joshi, B., Mishra, D., 2009. Study of pectinase production in submerged fermentation using different strains of *Aspergillus niger*. Int. J. Microbiol. Res. 1, 13–17.

Lee, C., Darah, I., Ibrahim, C.O., 2017. Efficiency of developed solid state bioreactor 'FERM-SOSTAT' on cellulolytic and xylanase enzymes production. Sains Malays. 46, 1249–1257.

Mahmoodi, M., Najafpour, G.D., Mohammadi, M., 2019. Bioconversion of agroindustrial wastes to pectinases enzyme via solid state fermentation in trays and rotating drum bioreactors. In: Biocatalysis and Agricultural Biotechnology, vol. 21, p. 101280.

Manohar, B., Divakar, S., 2005. An artificial neural network analysis of porcine pancreas lipase catalysed esterification of anthranilic acid with methanol. Process Biochem. 40, 3372–3376. https://doi.org/10.1016/j.procbio.2005.03.045.

Maryam, F., Mohammad, F., 2015. Production of proteases in a novel trickling tray bioreactor. Waste Biomass Valori. 6, 475–480. https://doi.org/10.1007/s12649-015-9371-6.

Mazumder, S., Basu, Soumendra, Mukherjee, M., 2009. Laccase production in solid-state and submerged fermentation by *Pleurotus ostreatus*. Eng. Life Sci. 9, 45–52. https://doi.org/10.1002/elsc.200700039.

Mitchell, D.A., Pitol, L.O., Biz, A., Finkler, A.T.J., de Lima Luz, L.F., Krieger, N., 2019. Design and operation of a pilot-scale packed-bed bioreactor for the production of enzymes by solid-state fermentation. In: Steudler, S., Werner, A., Cheng, J. (Eds.), Solid State Fermentation, Advances in Biochemical Engineering/Biotechnology, vol. 169. Springer, Cham. https://doi.org/10.1007/10_2019_90.

Mlook, W., 2015. Production optimization of pullulanase enzyme produced by *Bacillus cereus* isolated from Syrian sources. Int. Food Res. J. 22, 1824–1830.

Mouna, K., Hanen, B., Marwa, B.R., Yosri, B.N., Slim, T., Lotif, M., Radhouane, K., 2020. Overproduction of glucose oxidase by *Aspergillus tubingensis* CTM 507 randomly obtained mutants and study of its insecticidal activity against *Ephestia kuehniell*. BioMed Res. Int. https://doi.org/10.1155/2020/9716581. Article ID 9716581.

Nagata, Y., Chu, K.H., 2003. Optimization of a fermentation medium using neural networks and genetic algorithms. Biotechnol. Lett. 25, 1837–1842. https://doi.org/10.1023/A:1026225526558.

Nampoothiri, K.M., Tomes, G.J., Roopesh, K., George Szakacs, Viviana Nagy, Carlos Ricardo Soccol, Ashok Pankey, 2004. Thermostable phytase production by *Thermoascus aurantiacus* in submerged fermentation. Appl. Biochem. Biotechnol. 118, 205–214.

Ozkaya, B., Kaksonen, A.H., Sahinkaya, E., Puhakka, J.A., 2019. Fluidized bed bioreactor for multiple environmental engineering solutions. Water Res. 150, 452–465.

Pandey, A., Radhakrishnan, S., 1992. Packed-bed column bioreactor for production of enzyme. Enzym. Microb. Technol. 14, 486–488.

Parashar, S.K., Srivastava, S.K., Garlapati, V.K., Dutta, N.N., 2019. Production of microbial enzyme triacylglycerol acyl hydrolases by *Aspergillus sydowii* jpg01 in submerged fermentation using agro-residues. Asian J. Microbiol. Biotechnol. Environ. Sci. 21 (4), 1076–1079.

Perez, C.L., Casciatori, F.P., Thoméo, J.C., 2019. Strategies for scaling-up packed-bed bioreactors for solid-state fermentation: the case of cellulolytic enzymes production by a thermophilic fungus. Chem. Eng. J. 361, 1142–1151. https://doi.org/10.1016/j.cej.2018.12.169.

Perez, C.L., Casciatori, F.P., Thoméo, J.C., 2020. Improving enzyme production by solid-state cultivation in packed-bed bioreactors by changing bed porosity and airflow distribution. Bioproc. Biosyst. Eng. 44, 537−548. https://doi.org/10.1007/s00449-020-02466-7.

Ramos, O.S., Malcata, F.X., 2011. Industrial biotechnology and commodity products. In: Comprehensive Biotechnology, second ed.

Ravichandra, P., Subhakar, J.A., 2007. Alkaline protease production by submerged fermentation in stirred tank reactor using *Bacillus licheniformis* NCIM-2042: effect of aeration and agitation regimes. Biochem. Eng. J. 34, 185−192.

Reddy, E.R., Babu, R.S., durthi Chandrasai, P., Madhuri, P., 2017. Neural network modeling and genetic algorithm optimization strategy for the production of l-asparaginase from Novel Enterobacter sp. J. Pharmaceut. Sci. Res. 9, 124−130.

Rodrigues da Silva, R., Pereira de Freitas Cabral, T., Rodrigues, A., Cabral, H., 2013. Production and partial characterization of serine and metallo peptidases secreted by *Aspergillus fumigatus* Fresenius in submerged and solid state fermentation. Braz. J. Microbiol. 44, 235−243. https://doi.org/10.1590/S1517-83822013000100034.

Rodríguez-Durán, L., Contreras-Esquivel, J.C., Rodríguez, R., Prado-Barragán, A., Aguilar, C.N., 2011. Optimization of tannase production by *Aspergillus niger* in solid-state packed-bed bioreactor. J. Microbiol. Biotechnol. 21, 960−967.

Rodríguez-Jasso, R.M., Mussatto, S.I., Sepúlveda, L., Agrasar, A.T., Pastrana, L., Aguilar, C.N., Teixeira, J.A., 2013. Fungal fucoidanase production by solid-state fermentation in a rotating drum bioreactor using algal biomass as substrate. Food Bioprod. Process. 91, 587−594.

Romero Gomez, S., Augur, C., Viniegra-González, Gustavo, G., 2000. Invertase production by Aspergillus niger in submerged solid-state fermentation. Biotechnol. Lett. 22, 1255−1258. https://doi.org/10.1023/A:1005659217932.

Ruiz, H.A., Rodríguez-Jasso, R.M., Rodríguez, R., Contreras-Esquivel, J.C., Aguilar, C.N., 2012. Pectinase production from lemon peel pomace as support and carbon source in solid-state fermentation column-tray bioreactor. Biochem. Eng. J. 65, 90−95.

Satapute, P., Milan, V.K., Shivakantkumar, S.A., Jogaiah, S., 2019. Influence of triazole pesticides on tillage soil microbial populations and metabolic changes. Sci. Total Environ. 651, 2334−2344.

Sathyanarayana, N., Gummadi, Sunil Kumar, D., 2006. Enhanced production of pectin lyase and pectate lyase by *Debaryomyces nepalensis* in submerged fermentation by statistical methods. Am. J. Food Technol. 1, 19−33.

Sayyed, R.Z., Shimpi, G.B., Chincholkar, S.B., 2010. Constitutive production of extracellular glucose isomerase by an osmophillic *Aspergillus* sp. under submerged conditions. J. Food Sci. Technol. 47, 496−500. https://doi.org/10.1007/s13197-010-0084-3.

Shaikh, A., 2012. Optimization of Lactase production by submerged fermentation using *Aspergillus oryzae*. Int. J. Biol. Pharmaceut. Res. 3, 743−746.

Shindia, A.A., Khalaf, S.A., Yassin, M.A., 2013. Production and partial characterization of β-glucanase from *Aspergillus niger* JQ1516491 under submerged and solid state fermentation. Asian J. Microbiol. Biotechnol. Environ. Sci. 15, 459−472.

Songulashvili, G., Spindler, D., Jimenéz-Tobón, G.A., Jaspers, C., Kerns, G., Penninckx, M.J., 2015. Production of a high level of laccase by submerged fermentation at 120-L scale of *Cerrena unicolor* C-139 grown on wheat bran. Comptes Rendus Biol. 338, 121−125.

Teng, Y., Xu, Y., 2008. Culture condition improvement for whole-cell lipase production in submerged fermentation by Rhizopus chinensis using statistical method. Bioresour. Technol. 99, 3900−3907.

Tripathi, C., Mishra, S., 2009. Culture conditions for the production of a-galactosidase by Aspergillus parasiticus MTCC-2796: a novel source. Electron. J. Biotechnol. 12, 4–5. https://doi.org/10.4067/S0717-34582009000400004.

Vaseghi, Z., Najafpour, G.D., Mohseni, S., Mahjoub, S., 2012. Production of active lipase by Rhizopus oryzae from sugarcane bagasse: solid state fermentation in a tray bioreactor. Int. J. Food Sci. Technol. 1–7.

Zhang, L., Han, X., Du, M., Zhang, Y., Feng, Z., Yi, H., Zhang, Y., 2012. Enhancement of transglutaminase production in *Streptomyces mobaraensis* as achieved by treatment with excessive MgCl2. Appl. Microbiol. Biotechnol. 93, 2335–2343.

CHAPTER

Laboratory methods for enzyme assays

5

SUBCHAPTER

Assay of enzyme—fructosyltransferase

5.1

Before you begin

Timing: For fungi—4 days.
For bacteria or yeast—48 h.
Commercial enzyme preparation—2 h.

Reagent preparation

1. **Czapak Dox liquid medium (g/L):** Dissolve sucrose 30.0 g, $NaNO_3$ 3.0 g, $MgSO_4$ 0.5 g, KCl 0.5 g, $FeSO_4$ 0.01 g, KH_2PO_4 1.0 g, distilled water 1000 mL, and adjust the pH to 5.50.
2. **Sucrose solution (55%):** 55 g of sucrose is added to a volumetric flask, and distilled water is added in a quantity to dissolve the sugar first. Once the solution is clear, the volume is made up to the 100 mL mark.
3. **Sodium acetate buffer (0.1 M; pH 5.5):**
   a. **Solution A:** 1 M acetic acid was prepared by adding 12.06 mL of glacial acetic acid to 200 mL of distilled water.
   b. **Solution B:** 1 M NaOH was prepared by dissolving 4 g of NaOH in 100 mL of distilled water.

   146.0 mL of solution A and 100 mL of solution B were mixed together. The pH was adjusted with weak acid and alkali, and the final volume was made to 1000 mL with distilled water.
4. **Di-nitrosalicylic acid reagent (DNSA):**
   a. **Solution A:** 300 g of sodium potassium tartarate was dissolved in 500 mL of distilled water.

Protocols and Applications in Enzymology. https://doi.org/10.1016/B978-0-323-91268-6.00005-3

b. **Solution B:** 2 M sodium hydroxide solution was prepared by adding 16 g of sodium hydroxide to 200 mL of distilled water
c. **Solution C:** 10 g of 3,5-di-nitrosalicylic acid was dissolved in 200 mL of solution B. The reagent was prepared by mixing solutions A and C. The final volume was made up to 1 L with distilled water.

4. **Preparation of standard curve for reducing sugars:**
   a. **Stock solution of Glucose (0.1 g/mL):** Dissolve 1 g of glucose in 10 mL of distilled water using a volumetric flask.
   b. **Standard working solution (0.001 g/mL):** Take 1 mL (0.1 g) of stock glucose solution and make up the volume to 100 mL using a volumetric flask.

Key resources table

Reagent or resource	Source	Identifier
Crude enzyme prepared in laboratory Commercial enzyme preparation of fructosyltransferase		
Fungal culture		
Aspergillus niger Trichoderma species Any fungus showing fructosyltransferase production in primary screening	Isolated or Stock culture	

Materials and equipment

- **Materials:** volumetric flasks, conical flasks, funnel, test tubes, test tube stand, pipettes, and pipette stand.
- **Equipment:** weighing balance, autoclave, incubator, shaking water bath, boiling water bath, and visible spectrophotometer.
- **Alternatives**: colorimeter can be used instead of spectrophotometers.

Day 1—production of fructosyltransferase enzyme

1. **Preparation of Czapak Dox liquid medium (g/L)**: Prepare 100 mL medium in a 500 mL conical flask.
   All the ingredients except Agar—agar powder and KH2PO4 are to be dissolved in 500 mL of distilled water. KH2PO4 should be dissolved separately in a minimal amount of water and finally add to the solution containing other nutrients to avoid precipitation.

The pH of the medium should be adjusted to 5.50 by using 0.1 N HCl and 0.1 N NaOH. The volume was made up to 1 L with distilled water.
The medium should be autoclaved at 10 lbs/in.^{2} at 115°C for 20 min to avoid charring of carbohydrate content.

2. **Inoculation of the medium:**
The autoclaved medium is allowed to cool and inoculated with the selected culture of fungus under aseptic conditions. The medium is incubated at 27°C for 4 days.
DAY 4—After incubation, filter the fermented broth and use filtrate as a crude enzyme source.
Note: The crude enzyme should be stored at 4°C until further required in the protocol.

Preparation of standard curve for reducing sugars

Timing: 1 h

1. The different concentrations of working glucose solution were taken in eight test tubes as 0.1 mL, 0.2 mL, 0.3 mL, 0.4 mL, 0.5 mL, 0.6 mL, 0.7 mL, and 0.8 mL, respectively, along with a blank tube containing 1 mL of distilled water.
2. Two milliliters of DNSA was added to each tube, and all tubes were kept in a boiling water bath for 10 min. Total volume is made up to 10 mL with distilled water in all tubes. A spectrophotometer is set against a blank of 2 mL of DNSA, and distilled water is added to make up the volume to 10 mL. Read absorbance at 540 nm.
3. Standard graph is to be prepared by plotting optical density against the concentration of glucose in micromoles. Fig. 5.1: Standard curve of reducing sugar.

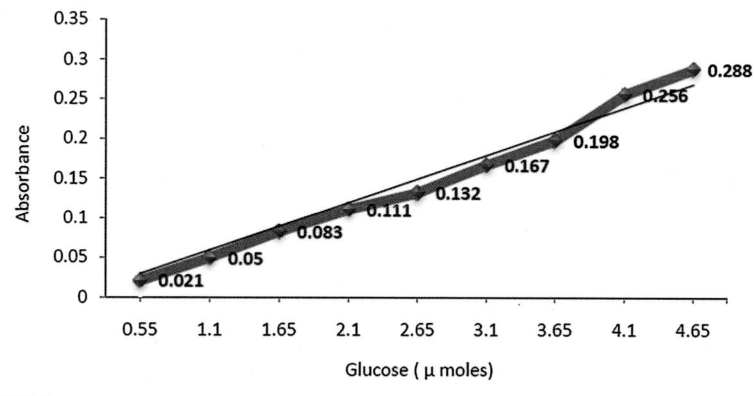

FIGURE 5.1

Standard curve of glucose (read absorbance at 540 nm).

Fructosyltransferase assay

1. Test sample: 0.5 mL of crude enzyme and 1.0 mL of buffer 55% sucrose solution is added to the reaction mixture and incubated at 37°C for 1 h.
2. Control: 1.0 mL of 55% sucrose solution is taken in a tube. 2 mL of DNSA reagent is added to it prior to the addition of 0.5 mL of the enzyme. The tube is incubated for 1 h
3. Blank: 55% sucrose solution reagent is added to the tube and incubated at 37°C for 1 h.
4. The enzymatic reaction is terminated by keeping all the test tubes at 100°C in a water bath for 10 min.
5. 2 mL DNSA is added to each tube except the control, and all tubes were kept in a boiling water bath for 10 min. After cooling, the final volume was made to 10 mL with distilled water in all tubes.
6. Absorbance is to be read at 540 nm in a spectrophotometer against a Blank. The standard graph is prepared by plotting optical density against the concentration of glucose (Fig. 5.1).
    1. A standard graph is plotted as shown in Fig. 5.1.

Note: Note down the absorbance of the control and experimental tubes using spectrophotometer and colorimeter. The experimental tubes are taken in triplicates for accuracy.

Quantification of the enzyme

Timing: 1 h

1. The quantity of enzyme is to be estimated by calculation.
    a. Standard curve is required for calculation.
    b. The standard curve reducing sugar concentration in micromoles to make the calculation easier.
    c. Convert the concentration of your stock solution (g/mL) to micromoles/mL.
    d. Now you get the concentration of glucose in each tube of the standard curve in micromoles, which is plotted on the X-axis (Fig. 5.1).
    e. Extrapolate the absorbance of each tube on X-axis to get the concentration of the product (glucose) formed.
    f. Divide the concentration derived by the incubation period.
    g. The value provides you the of IU in 0.5 mL. Multiply it with 2. The answer provides you that the quantity of enzyme present in 1 mL is referred as IU/mL.

Critical: The quantification of enzymes in any source is generally done in the International unit (IU).

One IU is defined as the quantity of enzyme required to release 1 micromole of product in 1 min.

As fructosyltransferase act upon its substrate and releases reducing sugars, the product is taken as glucose for calculation of IU.

Pause Point:

As this experiment involves production of enzyme by microbes its duration varies (for fungi—4 days, for bacteria or yeast—48 h).

1. Day 1: Reagent preparation, media preparation, and inoculation.
2. Day 2: Pause (bacteria or yeast), Day 2 and 3 for molds
3. Day 3: Performance of assay and calculation of IU/mL of fructosyltransferase. In the case of fungi (molds), this step is performed on Day 5.

Expected outcomes

1. This protocol efficiently delivers the amount of enzyme present in the given source. The quantification of the enzyme is further required for secondary screening and selection.

Quantification and statistical analysis

Once the IU/mL is calculated, the enzyme source present in the total volume of the given source can be easily calculated. As this protocol is a single-source estimation, comparative statistics cannot be applied. The triplicates of experimental are used to calculate the mean and the standard deviation to improve the efficacy of the process.

Advantages

1. It is a simple process.
2. Any enzyme source can be analyzed for enzyme quantification.
3. Reagents and equipment required are simple.
4. The assay procedure takes a short time.
5. Multiple enzyme sources can be handled at a time.

Limitations

The method is solely based on absorbance, hence instrumentation errors can be incorporated.

Absolute accuracy is required while additions of enzyme and substrate.

Optimization and troubleshooting

This is the most common problem encountered in fermentations that affect enzyme production.

Enzyme activity is not detected due to inhibitors produced by contaminants.

Medium precipitates after sterilization, and no growth occurs.

Medium is proper but no growth occurs.

All these reasons hamper enzyme production and no activity is registered in the crude sample.

Potential solution to optimize the procedure.

If contamination occurs, there is no alternative to discard and start fresh.

Strict sterilization and aseptic transfers are the measures that can avoid contaminations.

Di Potassium hydrogen phosphate has to be added as a solution after all the ingredients are dissolved in a partial volume of water.

The medium should be properly cooled before inoculation.

Many times the enzymes are sensitive to Fe salts. If all parameters are proper, then Fe salt can be avoided to cross-check to study its inhibitory effect on the enzyme.

Safety considerations and standards

Di-nitrosalicylic acid should be handled with care. It should not be allowed to come in contact with skin and eyes.

After the assay is complete, the fermentation remnants should be discarded after proper decontamination.

Alternative methods/procedures

Many authors have used other methods and compared the efficiency of reducing sugar estimation to the Millers method (Miller, 1959; Anupama et al., 2014). The DNS method is traditionally used for sugar assay although there are reports of reduced accuracy associated with this method (Bailey et al., 1992). The suggested alternative is NS reducing sugar method (Nelson, 1994; Somogyi, 1952). A microtitre plate assay method was suggested by Cairns (1987).

References

Anupama, N., Murali, M., Jogaiah, S., Amruthesh, K.N., 2014. Crude oligosaccharides from *Alternaria solani* with *Bacillus subtilis* enhance defensive activity against early blight disease of tomato. Asian J. Sci. Technol. 5 (7), 412–416.

Bailey, M.J., Biely, P., Poutanen, K., 1992. Interlaboratory testing of methods for assay of xylanase activity. J. Biotechnol. 23 (3), 257–270.
Cairns, A.J., 1987. Colorimetric microtiter plate assay of glucose and fructose by enzyme-linked formazan production: applicability to the measurement of fructosyl transferase activity in higher plants. Anal. Biochem. 167 (2), 270–278.
Miller, G.L., 1959. Use of dinitrosalicylic acid reagent for determination of reducing sugar. Anal. chem. 31, 426–428.
Nelson, N., 1944. A photometric adaption of the Somogyi method for the determination of glucose. J. Biol. Chem. 153, 375–380.
Somogyi, M.J., 1952. Notes on sugar determination. J. Biolol. Chem. 195, 19–23.

SUBCHAPTER

Assay of enzyme—Lipase 5.2

Before you begin

Timing: For fungi—4 days.
For bacteria or yeast—48 h.
Commercial enzyme preparation—2 h.

Reagent preparation

Nutrient medium for lipase production: Peptone—5 g, beef extract—3 g, NaCl—5 g, olive oil—1 mL, distilled water—1000 mL.

Phenolphthalein indicator: Dissolve 0.5 g of phenolphthalein powder to 50 mL of 95% ethanol and stir well. Now, make up the volume to 100 mL by adding distilled water.

Sodium hydroxide solution: Prepare 100 mL in distilled water using sodium hydroxide.

Lipase assay

200 mM Tris HCl buffer, pH 7.0: Dissolve 121.14 g Tris in 800 mL of H_2O.

Adjust pH to 7.0 with the appropriate volume of concentrated HCL. Bring the final volume to 1 L with deionized water.

Olive oil substrate solution.

95% Ethanol.

0.9% Phenolphthalein indicator solution (dissolve 0.5 g phenolphthalein powder to 50 mL of 95% ethanol and stir well. Now, make up the volume to 100 mL by adding distilled water).

50 mM Sodium hydroxide solution—standardized (NaOH) (prepare 100 mL in deionized water using sodium hydroxide).

Lipase crude enzyme solution/commercial source.

Key resources table

Reagent or resource	Source	Identifier
Crude enzyme prepared in the laboratory		
Commercial enzyme preparation of lipase		
Fungal culture		
Aspergillus niger	Isolated or stock culture	
Any fungus showing lipase production in primary screening		

Materials and equipment

Materials: Volumetric flasks, conical flasks, funnel, test tubes, test tube stand, pipettes, and pipette stand and burette.

Equipment: Weighing balance, autoclave, incubator, shaking water bath, and boiling water bath.

Day 1—production of lipase enzyme

1. Preparation of nutrient medium (g/L): Prepare 100 mL of medium in a 500 mL conical flask.
2. All the ingredients were dissolved in 100 mL of distilled water.
3. The pH of the medium should be adjusted to 5.50 by using 0.1N HCl and 0.1N NaOH. The volume was made up to 1 L with distilled water.
4. The medium should be autoclaved at 10 lbs/in.^{2} at 115°C for 20 min.

Inoculation of the medium

The autoclaved medium is allowed to cool and inoculated with the selected culture of fungus under aseptic conditions. The medium is to be incubated at 27°C for 4 days (fungi) or 37°C for 24 h (bacteria).

Day 2/Day 5—After incubation, filter the fermented broth and use filtrate as crude enzyme source.

Note: The crude enzyme should be stored at 4°C until further required in the protocol.

Assay procedure: Lipase enzyme (titrimetric method)

Test sample: Pipette deionized water 2.50 mL.

Add buffer 1.00 mL and olive oil 3.00 mL in a conical flask.

Mix by swirling and equilibrate to 37°C. Then, add reagent 1 mL of enzyme solution.

Mix by swirling and incubate at 37°C for exactly 30 min.

Add 95% of ethanol 3.00 mL.

Mix by swirling and then add 4 drops of phenolphthalein.

Titrate the solution with NaOH to a light pink color.

Blank sample.

Pipette deionized water 2.50 mL.

Add buffer 1.00 mL and olive oil 3.00 mL in a conical flask.

Mix by swirling and equilibrate to 37°C. Then, add reagent 1 mL of distilled water.

Mix by swirling and incubate at 37°C for exactly 30 min.

Add 95% of ethanol 3.00 mL.

Mix by swirling and then add 4 drops of phenolphthalein.

Titrate the solution with NaOH to a light pink color.

Calculations:

$$\text{Units/mL enzyme} = (\text{NaOH})\ (\text{Molarity of NaOH})\ (1000)\ (2)\ (\text{df}) \quad (5.1)$$

where

(NaOH) = Volume (in milliliters) of reagent used for test minus volume (in milliliters) of reagent used for blank.

1000 = Conversion factor from milliequivalent to microequivalent

2 = Time conversion factor from 30 min to 1 h

df = Dilution factor 1 = Volume (in milliliter) of enzyme used

Critical: The quantification of enzymes in any source is generally done in the International unit (IU).

One IU of lipase is defined as the quantity of enzyme required to hydrolyze 1 micromole equivalent of fatty acid in 1 h.

Pause point:

As this experiment involves the production of enzymes by microbes its duration varies (for fungi—4 days, for bacteria or yeast—48 h).

1. Day 1: Reagent preparation, media preparation, and inoculation,
2. Day 2: Pause (bacteria or yeast), Day 2 and 3 for molds
3. Day 3: Performance of assay and calculation of IU/mL of lipase. In the case of fungi (molds), this step is performed on Day 5.

Expected outcomes

1. This protocol efficiently delivers the amount of enzyme present in the given source. The quantification of the enzyme is further required for secondary screening and selection.

Quantification and statistical analysis

Once the IU/mL is calculated, the enzyme source present in the total volume of the given source can be easily calculated. As this protocol is a single-source estimation, comparative statistics cannot be applied. The triplicates of experimental are used to calculate the mean and the standard deviation to improve the efficacy of the process.

Advantages

1. It is a simple process.
2. Any enzyme source can be analyzed for enzyme quantification.
3. Reagents and equipment required are simple.
4. The assay procedure takes a short time.
5. Multiple enzyme sources can be handled at a time.

Limitations

The method is solely based on titration.

Absolute accuracy is required while additions of enzyme and substrate.

Optimization and troubleshooting

This is the most common problem encountered in fermentations that affect enzyme production.

Enzyme activity is not detected due to inhibitors produced by contaminants.

Medium precipitates after sterilization, and no growth occurs.

Medium is proper but no growth occurs.

All these reasons hamper enzyme production and no activity is registered in the crude sample.

Potential solution to optimize the procedure.

If contamination occurs, there is no alternative to discard and start fresh.

Strict sterilization and aseptic transfers are the measures that can avoid contaminations.

Di Potassium hydrogen phosphate has to be added as a solution after all the ingredients are dissolved in a partial volume of water.

The medium should be properly cooled before inoculation.

Many times the enzymes are sensitive to Fe salts. If all parameters are proper, Fe salt can be avoided to cross-check to study its inhibitory effect on the enzyme.

Safety considerations and standards

Maintenance of pH and temperature is critical for the experiment.

After the assay is complete the fermentation remnants should be discarded after proper decontamination.

Alternative methods/procedures

The above method is a volumetric standard protocol for lipase assay (Cherry and Grandall, 1932). Various instrument based assay methods are now available (Stoytcheva et al., 2012).

References

Cherry, I.S., Crandall, L.A., 1932. The specificity of pancreatic lipase; its appearance in the blood after pancreatic injury. Am. J. Physiol. 100, 266–273.

Stoytcheva, M., Gisela, M., Roumen, Z., Jose, L., Velizar, G., 2012. Analytical methods for lipases activity determination: a review. Curr. Anal. Chem. 8, 400–407. https://doi.org/10.2174/157341112801264879.

SUBCHAPTER

Assay of enzyme—protease 5.3

Before you begin

Timing: For fungi—4 days.
For bacteria or yeast—48 h.
Commercial enzyme preparation—2 h.

Reagent preparation

4. **Caesin broth medium (g/L):** Dissolve starch 10.0 g, casein 0.30 g, KNO_3 2.0 g $MgSO_4$ 0.05 g, NaCl 2.0 g, $CaCO_3$, $FeSO_4$ 0.01 g, K2HPO4 2.0 g, distilled water 1000 mL, and adjust the pH to 7.0.
5. 0.65% Casein was used as substrate. It is prepared from alkali-soluble casein, which was dissolved in 0.1M phosphate buffer (7.0). The insoluble portion is dissolved by addition of the alkali. The pH is adjusted to 8.0 with 0.1 M sodium hydroxide.
6. **Phosphate buffer (0.1 M; pH 7.0):**
   a. **Solution A:** Prepare sodium phosphate dibasic stock Na2HPO4 (0.5 M) by dissolving 35.5 g of sodium phosphate dibasic in a final volume of 500 mL of H_2O. Measure 80 mL and let the volume made up to 400 mL. This is a 0.1 M solution.
   b. **Solution B:** Prepare sodium phosphate monobasic stock NaH2PO4 (0.5 M) by dissolving 30 g of anhydrous sodium phosphate monobasic in a final volume of 500 mL of H_2O. Take 30 mL and make up the volume to 150 mL. This is a 0.1 M solution.

   Take 400 mL of Solution A (0.1M) and add solution B till the pH is maintained at 7.0.
7. Trichloroacetic acid reagent: Dilute TCA in the ratio of 1:55 with distilled water.
8. Folin-Ciocalteu's phenol reagent (FC reagent): The Fc (2M) reagent available commercially is to be diluted in ratio of 1:4.
9. Sodium carbonate solution: 500 mM solution is prepared by dissolving 53 mg of anhydrous Na2CO3/mL of distilled water.
4. **Preparation of standard curve for L-Tyrosine:**
   Standard tyrosine solution: Prepare 0.2 mg/mL of L-Tyrosine solution. Gentle heating is suggested for complete solubilization.

Key resources table

Reagent or resource	Source	Identifier
Crude enzyme prepared in the laboratory Commercial enzyme preparation of protease		
Fungal culture		
Aspergillus niger Trichoderma species Any fungus showing xylanase production in primary screening	Isolated or stock culture	

Materials and equipment

- **Materials:** Volumetric flasks, conical flasks, funnel, test tubes, test tube stand, pipettes, and pipette stand.
- **Equipment:** Weighing balance, autoclave, incubator, shaking water bath, boiling water bath, and visible spectrophotometer.
- **Alternatives**: Colorimeter can be used instead of spectrophotometers.

Day 1—production of protease

1. **Preparation of Casein broth medium (g/L)**: Prepare 100 mL medium in a 500 mL conical flask.
   All the ingredients except KH2PO4 were dissolved in 500 mL of distilled water.
   KH2PO4 was dissolved separately in minimal amount of water and finally added to solution containing other nutrients to avoid precipitation.
   The pH of the medium was adjusted to 7.0 by using 0.1N HCl and 0.1N NaOH.
   The volume was made up to 1 L with distilled water.
   The medium was autoclaved at 15 lbs/in.^{2} at 121°C for 15 min to avoid charring of carbohydrate content.
2. **Inoculation of the medium:**
   The autoclaved medium is allowed to cool and inoculated with the selected culture of fungus/bacteria/yeast under aseptic conditions. The medium is incubated at 27°C for 4 days.
   Day 4—After incubation, filter the fermented broth and use filtrate as a crude enzyme source.
   Note: The crude enzyme should be stored at 4°C until further required in the protocol.

Preparation of standard curve for L-Tyrosin

Timing: 1 h

1. The different concentrations of standard L-Tyrosine are taken in eight test tubes as 0.05, 0.1, 0.2, 0.4, and 0.5 mL, respectively, along with a blank tube containing 1 mL of distilled water.
2. Add water to make up the volume to 1 mL.
3. Add 5 mL of sodium carbonate solution to each tube.
4. Immediately add 1 mL of FC reagent to each tube.
5. Mix properly and allow it to incubate for 30 min at 37°C.
6. Read Absorbance at 660 nm.
7. Standard graph was prepared by plotting optical density against the concentration of L-Tyrosine in micromoles. Fig. 5.2: Standard curve of reducing sugar.

Protease assay:

1. Take 4 test tubes and label them as control and experimental (E1, E2, E3).
2. Add 5 mL 0.65% casein in 0.1 M phosphate buffer (pH 7.0) is used as substrate.
3. Add 3 mL of trichloroacetic acid to the control tube.
3. The filtrate is used as crude enzyme source. 1 mL of the crude enzyme is added and incubated at 50°C for 60 min.
4. 3 mL of trichloroacetic acid is used to terminate the reaction. in experimental tubes.
5. Centrifuge the reaction mixture at 5000 rpm for 15 min.
6. Withdraw 0.5 mL of supernatant in another tube and add 2.5 mL of sodium carbonate.
7. Incubate the mixture for 20 min and add 0.5 mL of FC reagent.
8. Read the absorbance at 660 nm using a spectrophotometer.

A standard graph is plotted as shown in Fig. 5.2.

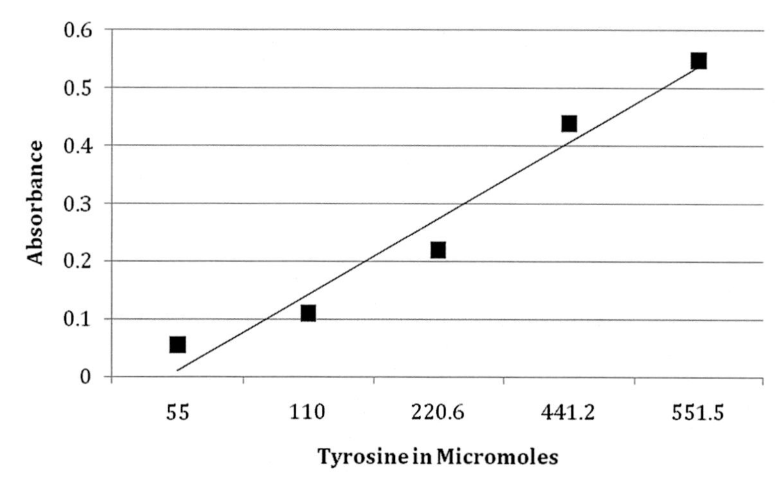

FIGURE 5.2

Standard curve of Tyrosine.

Note: Note down the absorbance of the control and experimental tubes using spectrophotometer and colorimeter The experimental tubes are taken in triplicates for accuracy.

Quantification of the enzyme

Timing: 1 h

1. The quantity of enzyme is to be estimated by calculation.
   a. Standard curve is required for calculation.
   b. The standard of tyrosine is in micromoles to make the calculation easier.
   c. Convert the concentration of your stock solution (mg/mL) to micromoles/mL.
   d. Now you get the concentration of Tyrosine in each tube of the standard curve in micromoles, which is plotted on the X-axis (Fig. 5.2).
   e. Extrapolate the absorbance of each tube on the X-axis to get the concentration of the product (glucose) formed.
   f. Divide the concentration derived by the incubation period (60 min).
   g. The value provides you the IU in per mL. The answer provides you that the quantity of enzyme present in 1 mL is referred as IU/mL.

Critical: The quantification of enzymes in any source is generally done in International unit (IU).

One IU is defined as the quantity of enzyme required to release 1 micromole of Tyrosine in 1 min.

As protease acts upon its substrate and releases amino acids, the product is taken as Tyrosine released for calculation of IU.

Pause point:

As this experiment involves the production of enzymes by microbes, its duration varies (for fungi—4 days, for bacteria or yeast—48 h).

1. Day 1: Reagent preparation, media preparation, and inoculation,
2. Day 2: Pause (bacteria or yeast), Day 2 and 3 for molds.
3. Day 3: Performance of assay and calculation of IU/mL of xylanase. In the case of fungi (molds), this step is performed on Day 5.

Expected outcomes

1. This protocol efficiently delivers the amount of enzyme present in the given source. The quantification of the enzyme is further required for secondary screening and selection.

Quantification and statistical analysis

Once the IU/mL is calculated, the enzyme source present in the total volume of the given source can be easily calculated. As this protocol is a single-source estimation, comparative statistics cannot be applied. The triplicates of experimental are used to calculate the Mean and the standard deviation to improve the efficacy of the process.

Advantages

1. It is a simple process.
2. Any enzyme source can be analyzed for enzyme quantification.
3. Reagents and equipment required are simple.
4. The assay procedure takes a short time.
5. Multiple enzyme sources can be handled at a time.

Limitations

The method is solely based on absorbance, hence instrumentation errors can be incorporated.

Absolute accuracy is required while additions of enzyme and substrate.

Optimization and troubleshooting

This is the most common problem encountered in fermentations that affect enzyme production.

Enzyme activity is not detected due to inhibitors produced by contaminants.

Medium precipitates after sterilization, and no growth occurs.

Medium is proper but no growth occurs.

All these reasons hamper enzyme production and no activity is registered in the crude sample.

Potential Solution to optimize the procedure.

If contamination occurs, there is no alternative to discard and start fresh.

Strict sterilization and aseptic transfers are the measures that can avoid contaminations.

Di Potassium hydrogen phosphate has to be added as a solution after all the ingredients are dissolved in a partial volume of water.

The medium should be properly cooled before inoculation.

Many times, the enzymes are sensitive to Fe salts. If all parameters are proper, then Fe salt can be avoided to cross-check to study its inhibitory effect on the enzyme.

Safety considerations and standards

Reagents should be handled with care.

After the assay is complete, the fermentation residue should be discarded after proper decontamination.

Alternative methods/procedures

This is a standard protocol for protease assay.

Further reading

Anson, M.L., 1938. J. Gen. Physiol. 22, 79–89.
Folin, O., Ciocalteau, V., 1929. J. Biol. Chem. 73, 627.

SUBCHAPTER

Assay of enzyme—xylanases

5.4

Before you begin

Timing: For fungi—4 days.
For bacteria or yeast—48 h.
Commercial enzyme preparation—2h.

Reagent preparation

10. **Czapak Dox liquid medium (g/L):** Dissolve sucrose 30.0 g, $NaNO_3$ 3.0 g, $MgSO_4$ 0.5 g, KCl 0.5 g, $FeSO_4$ 0.01 g, KH2PO4 1.0 g, Birchwood Xylan 5 g, distilled water 1000 mL, and adjust the pH to 5.50.
11. **Xylan solution (1%):** 1 g of xylan was dissolved in 100 mL of distilled water.
12. **Sodium acetate buffer (0.1M; pH 5.5):**
    a. **Solution A:** 1 M acetic acid was prepared by adding 12.06 mL of glacial acetic acid to 200 mL of distilled water.
    b. **Solution B:** 1 M NaOH was prepared by dissolving 4 g of NaOH in 100 mL of distilled water.

    146.0 mL of solution A and 100 mL of solution B were mixed together. The pH was adjusted with weak acid and alkali, and the final volume was made to 1000 mL with distilled water.
4. **Di-nitro salicylic acid reagent (DNSA):**
    a. **Solution A:** 300 g of sodium potassium tartarate was dissolved in 500 mL of distilled water.
    b. **Solution B:** 2 M sodium hydroxide solution was prepared by adding 16 g of sodium hydroxide to 200 mL of distilled water
    c. **Solution C:** 10 g of 3, 5-di-nitrosalicylic acid was dissolved in 200 mL of solution B. The reagent was prepared by mixing solutions A and C. The final volume was made up to 1 L with distilled water.
4. **Preparation of standard curve for reducing sugars:**
    a. **Stock solution of glucose (0.1 g/mL):** Dissolve 1 g of glucose in 10 mL of distilled water using a volumetric flask.
    b. **Standard working solution (0.001 g/mL):** Take 1 mL (0.1 g) of stock glucose solution and make up the volume to 100 mL using a volumetric flask.

Key resources table

Reagent or resource	Source	Identifier
Crude enzyme prepared in laboratory Commercial enzyme preparation of xylanase		
Fungal culture		
Aspergillus niger Trichoderma species Any fungus showing xylanase production in primary screening	Isolated or stock culture	

Materials and equipment

- **Materials:** Volumetric flasks, conical flasks, funnel, test tubes, test tube stand, pipettes, and pipette stand.
- **Equipment:** Weighing balance, autoclave, incubator, shaking water bath, boiling water bath, and visible spectrophotometer.
- **Alternatives**: Colorimeter can be used instead of spectrophotometers.

Day 1—production of xylanase enzyme

1. **Preparation of Czapak Dox liquid medium (g/L)**: Prepare 100 mL medium in 500 mL of conical flask.
   All the ingredients except Agar—agar powder and KH_2PO_4 were dissolved in 500 mL of distilled water. KH_2PO_4 was dissolved separately in a minimal amount of water and finally added to a solution containing other nutrients to avoid precipitation.
   The pH of the medium was adjusted to 5.50 by using 0.1N HCl and 0.1N NaOH. The volume was made up to 1 L with distilled water.
   The medium was autoclaved at 10 lbs/in.^{2} at 115°C for 20 min to avoid charring of carbohydrate content.
2. **Inoculation of the medium:**
   The autoclaved medium is allowed to cool and inoculated with the selected culture of fungus under aseptic conditions. The medium is incubated at 27°C for 4 days.
   Day 4—After incubation, filter the fermented broth and use filtrate as a crude enzyme source.
   Note: The crude enzyme should be stored at 4°C until further required in the protocol.

Preparation of standard curve for reducing sugars

Timing: 30 min–1 h

1. The different concentrations of working glucose solution were taken in eight test tubes as 0.1 mL, 0.2 mL, 0.3 mL, 0.4 mL, 0.5 mL, 0.6 mL, 0.7 mL, and 0.8 mL, respectively, along with a blank tube containing 1 mL distilled water.
2. 2 mL of DNSA was added to each tube, and all tubes were kept in a boiling water bath for 10 min. Total volume is made up to 10 mL with distilled water in all tubes. A spectrophotometer is set against a blank of 2 mL DNSA and distilled water added to make up the volume to 10 mL. Read absorbance at 540 nm.
3. Standard graph was prepared by plotting optical density against the concentration of glucose in micromoles. Fig. 5.1: Standard curve of reducing sugar

Xylanase assay

1. Test sample: 0.5 mL of crude enzyme and 1.0 mL of buffer 1% xylan solution reagent was added to the reaction mixture and incubated at 37°C for 1 h.
2. Control: 1.0 mL of 1% xylan solution was taken in a tube. 2 mL of DNSA reagent was added to it prior to addition of 0.5 mL of the enzyme. The tube was incubated for 1 h
3. Blank: 1% xylan solution reagent was added to the tube and incubated at 37°C for 1 h.
4. The enzymatic reaction was terminated by keeping all the test tubes at 100°C in a water bath for 10 min.
5. 2 mL of DNSA was added to each tube except the control, and all tubes were kept in boiling water bath for 10 min. After cooling, the final volume was made to 10 mL with distilled water in all tubes.
6. Absorbance was read at 540 nm in a spectrophotometer against a blank. The standard graph was prepared by plotting optical density against the concentration of glucose (Fig. 5.1).
   2. A standard graph is plotted as shown in Fig. 5.1.

Note: Note down the absorbance of the control and experimental tubes using spectrophotometer and colorimeter. The experimental tubes are taken in triplicates for accuracy.

Quantification of the enzyme

Timing: 1 h

1. The quantity of enzyme is to be estimated by calculation.
   a. Standard curve is required for calculation.
   b. The standard curve reducing sugar concentration in micromoles to make the calculation easier.
   c. Convert the concentration of your stock solution (g/mL) to micromoles/mL.

d. Now you get the concentration of glucose in each tube of the standard curve in micromoles, which is plotted on the X-axis (Fig. 5.1).
e. Extrapolate the absorbance of each tube on the X-axis to get the concentration of the product (glucose) formed.
f. Divide the concentration derived by the incubation period.
g. The value provides you the IU in 0.5 mL, multiply it with 2. The answer provides you the quantity of enzyme present in 1 mL referred as IU/mL.

Critical: The quantification of enzymes in any source is generally done in the International unit (IU).

One IU is defined as the quantity of enzyme required to release 1 micromole of product in 1 min.

As xylanase acts upon its substrate and releases reducing sugars, the product is taken as glucose for calculation of IU.

Pause point:

As this experiment involves the production of enzymes by microbes, its duration varies (for fungi—4 days, for bacteria or yeast—48 h)

1. Day 1: Reagent preparation, media preparation, and inoculation.
2. Day 2: Pause (bacteria or yeast).
3. Day 3: Performance of assay and calculation of IU/mL of Xylanase. In case of Fungi (molds) this step is performed on Day 5.

Expected outcomes

1. This protocol efficiently delivers the amount of enzyme present in the given source. The quantification of the enzyme is further required for secondary screening and selection.

Quantification and statistical analysis

Once the IU/mL is calculated, the enzyme source present in the total volume of the given source can be easily calculated. As this protocol is a single-source estimation hence comparative statistics cannot be applied. The triplicates of experimental are used to calculate the mean and the standard deviation to improve the efficacy of the process.

Advantages

1. It is a simple process.
2. Any enzyme source can be analyzed for enzyme quantification.
3. Reagents and equipment required are simple.

4. The assay procedure takes a short time.
5. Multiple enzyme sources can be handled at a time.

Limitations

The method is solely based on absorbance, hence instrumentation errors can be incorporated.

Absolute accuracy is required while additions of enzyme and substrate.

Optimization and troubleshooting

This is the most common problem encountered in fermentations that affect enzyme production.

Enzyme activity is not detected due to inhibitors produced by contaminants.

Medium precipitates after sterilization, and no growth occurs.

Medium is proper but no growth occurs.

All these reasons hamper enzyme production, and no activity is registered in the crude sample.

Potential solution to optimize the procedure.

If contamination occurs, there is no alternative to discard and start fresh.

Strict sterilization and aseptic transfers are the measures that can avoid contaminations.

Di Potassium hydrogen phosphate has to be added as solution after all the ingredients are dissolved in a partial volume of water.

The medium should be properly cooled before inoculation.

Many times, the enzymes are sensitive to Fe salts. If all parameters are proper, then Fe salt can be avoided to cross-check to study its inhibitory effect on the enzyme.

Safety considerations and standards

Di-nitrosalicylic acid should be handled with care. It should not be allowed to come in contact with skin and eyes.

After the assay is complete, the fermentation remnants should be discarded after proper decontamination.

Alternative methods/procedures

Many authors have used other methods and compared the efficiency of reducing sugar estimation to the Millers method (Miller, 1959). The DNS method is traditionally used for sugar assay although there are reports of reduced accuracy associated with this method (Bailey et al., 1992). The suggested alternative is NS reducing sugar method (Nelson, 1994; Somogyi, 1952; McClear and McGeough; 2015).

References

Bailey, M.J., Biely, P., Poutanen, K., 1992. Interlaboratory testing of methods for assay of xylanase activity. J. Biotechnol. 23 (3), 257–270.

McCleary, B.V., McGeough, P.A., 2015. Comparison of polysaccharide substrates and reducing sugar methods for the measurement of *endo*-1,4-β-Xylanase. Appl. Biochem. Biotechnol. 177, 1152–1163. https://doi.org/10.1007/s12010-015-1803-z.

Miller, G.L., 1959. Use of dinitrosalicylic acid reagent for determination of reducing sugar. Anal. chem. 31, 426–428.

Nelson, N., 1944. A photometric adaption of the Somogyi method for the determination of glucose. J. Biol. Chem. 153, 375–380.

Somogyi, M.J., 1952. Notes on sugar determination. J. Biol. Chem. 195, 19–23.

CHAPTER

Structure and functions of enzyme kinetics

6

6.1 Introduction

To develop a first clear understanding of enzyme kinetics, the chemical reaction can be taken as an example. In chemical reactions, the reaction proceeds in the forward direction where reactants are converted to product till a point is reached where equilibrium is established between the reactant and the product. Hence, the equilibrium state of any reaction is defined as a condition in which the reactants and products obtained have attained the level of concentration with no further tendency of interchange with respect to time. The properties of such a reaction system do not tend to change further. Following Fig. 6.1A and B helps to understand the equilibrium state of any reaction mixture under study. The figure explains that at any time 0 or at the starting point of the reaction, the concentration of the reactant is maximum and the concentration of the product is zero or at the start point of the reaction. With timescale, the reactant's concentration decreases and the concentration of the product increases till they reach the equilibrium.

In case of enzyme-catalyzed reaction, the intervention of enzyme molecule in a biological reaction alters the rate of reaction (increases or decreases). The fundamental principle of enzyme-catalyzed reaction lies in the fact that neither the concentration of enzyme nor their presence alter the equilibrium of any reaction. On the contrary, it helps to attain the equilibrium faster. The enzymes have no influence on the structure of reactants/products nor does it affects the energy status of the two species.

The equilibrium of most of the reactions is at a point where most of the reactants have been converted into product. Hence, the concentration of the reaction is negligible and product is at maximum concentration. Such equilibrium exists at far right part of the total reaction time.

$$\text{Reactants} \rightarrow \text{Products}$$

The equilibrium can be better understood if the reaction is rewritten as follows:

$$\text{Reactants} \rightleftarrows \text{Products}$$

Protocols and Applications in Enzymology. https://doi.org/10.1016/B978-0-323-91268-6.00001-6

(A)

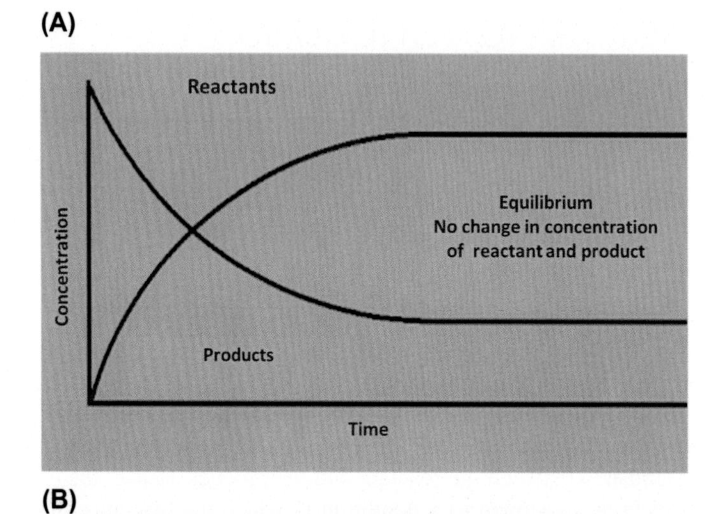

(B)

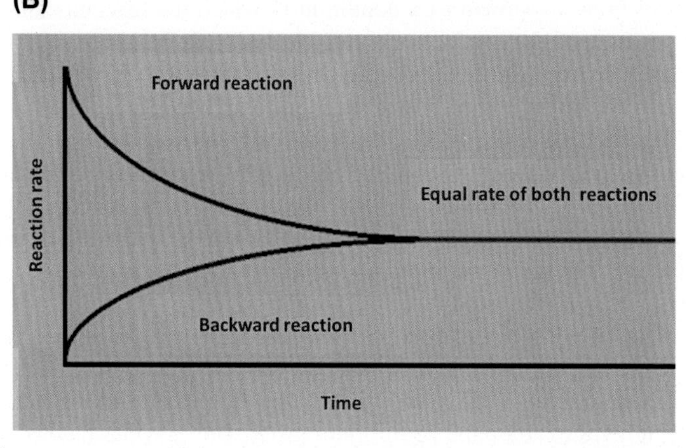

FIGURE 6.1

(A) concentration of substrate and product at equilibrium state. (B) Rate of forward reaction and backward reaction.

Thus, enzyme kinetics is related to study of rate of enzyme-catalyzed reaction. The enzyme kinetics deals with the role of enzymes in increasing the rate of reaction without altering the Gibb's energy change of the reaction. The enzyme kinetics is a sensitive field of enzymology, and reaction rate is drastically affected by various parameters like pH, temperature, and other environmental factors. The enzymes due to their biological nature are influenced positively or negatively by certain molecules termed as "activators or inhibitors" (Jogaiah et al., 2020).

6.2 Gibb's free energy and enzyme kinetic

The energy that is available for conversion of reaction to product is referred as Gibb's free energy till equilibrium is attained. The preliminary step for any reaction is the rate of collision of substrate molecule. If the collision rate is increased, it results in the increase in the rate of reaction. Following three factors are of utmost importance that influences the rate of the reaction.

6.2.1 Successful collision

For any collision to become successful, it must have colliding molecules with free energy equivalent to the activation energy. This is the minimum energy barrier that should be crossed over. Any reaction passes through an unstable intermediate state, which then separates to a more stable molecule as compared to the substrate.

6.2.2 The activation energy

The minimum energy requirement of any reaction that a substrate should possess to cross the barrier for getting transformed into the product. It is generally designated as Ea

6.3 The catalyst

The rate of the reaction is drastically affected by the presence of the catalyst. The catalyst is defined as a molecule that enhances the rate of reaction and it remains unaltered. The catalyst thus lowers the activation energy barrier, forming a transition complex at lower energy levels.

The enzyme kinetics is also dependent on the "law of mass action." Various factors influence the rate of reaction. When the reaction rate is proportionally dependent on the concentration of one substrate, it is called as the first-order reaction. Steady-state kinetics has been successfully used to study substrate affinity and the initial velocity of first-order reaction catalyzed by beta-galactosidase enzyme (Walsh et al., 2010).

If the reaction rate is dependent on concentration of two reactants or the second power of one reactant, then it is called as second-order reaction. When the reaction rate is independent of concentration of any reactant, it is called as zero-order reaction.

6.4 Enzyme kinetics and features of ECR

The prime features considered for enzyme catalyzed reactions are as follows:

A single substrate is involved in the enzyme catalyzed reactions yielding a single product (Sudisha et al., 2011).

The reaction is considered to be reversible and has rates of forward and backward reactions.

The reaction can be represented as follows:

$$E + S \rightleftarrows E + P$$

The rate of forward reaction is dependent on the rate of substrate used and the rate at which the product is formed. If both the rate of utilization of substrate and formation of product is proportional, the stoichiometric relationship can be represented as follows:

$$E + S \rightarrow E + P$$

After a fixed time interval, the reaction rate v will be

$$V = \frac{-d[S]}{dt} = \frac{+d[P]}{dt}$$

where v is the velocity or the reaction rate.

According to the law of mass action, the first-order reaction can also be represented as

$$V = \frac{-d[S]}{dt} = \frac{+d[P]}{dt} = k[S](\text{In case of first order reaction})$$

$$\text{Case of second order } V = -\frac{d[S_1]}{dt} = -\frac{d[S_2]}{dt} = +\frac{d[p]}{dt} = k\frac{[S_1]}{[S_2]}$$

6.5 The transition complex in enzyme-catalyzed reaction

As discussed earlier, the enzyme-catalyzed reaction proceeds in the forward direction with formation of ES complex. This complex is the transition complex with very subdued activation energy requirements as compared to uncatalyzed reactions. The free energy changes along the enzyme-catalyzed reaction are given in Fig. 6.2.

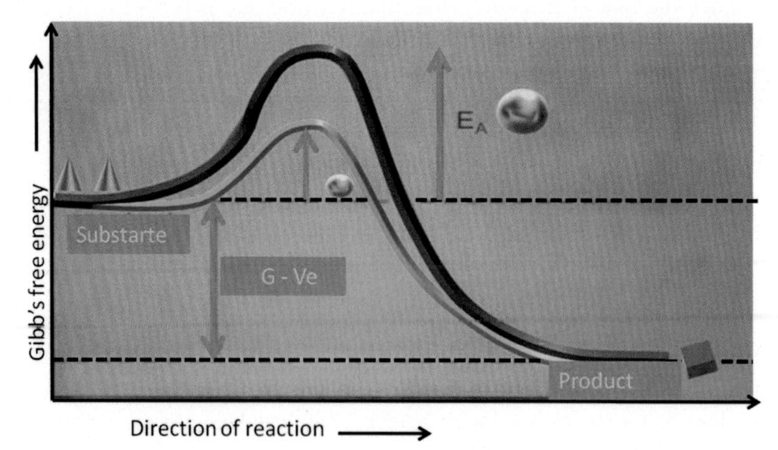

FIGURE 6.2

Free energy changes involved in enzyme catalyzed reactions.

6.6 The Michaelis–Menten equation

In biological systems, there are many instances that exemplify zero-order reaction. The reaction moves in forward direction independent of the concentration of the reactants. In case of single substrate reaction, if the concentration of substrate is in excess the reaction behaves as zero-order reaction and at limited substrate behaves as first-order reaction.

The enzyme catalyzed reaction for single substrate has a relationship between initial velocity and initial substrate concentration and represented in Fig. 9.3.

The equation that explains Fig. 6.3 is

$$V_0 = \frac{a[S]_0}{[S]_0 + b}$$

where a and b represent maximum initial velocity (V_{max}) and initial substrate concentration at which V_{max} was half.

As the concentration of substrate increases at constant enzyme concentration, the velocity increases. More substrate are bound to the active site, and more product are formed. This increase in velocity is proportional to the increase in the substrate concentration up to a point after which enzyme saturation is seen, and V_{max} is achieved. Beyond the V_{max} if substrate concentration is increased further, there is no influence on the enzyme rate. The velocity of reaction at this point is represented as follows.

$$V_{max} = K_2[E]$$

At constant enzyme concentration, the substrate concentration governs the rate of formation of enzyme substrate complex. Alternatively, if analysis of product

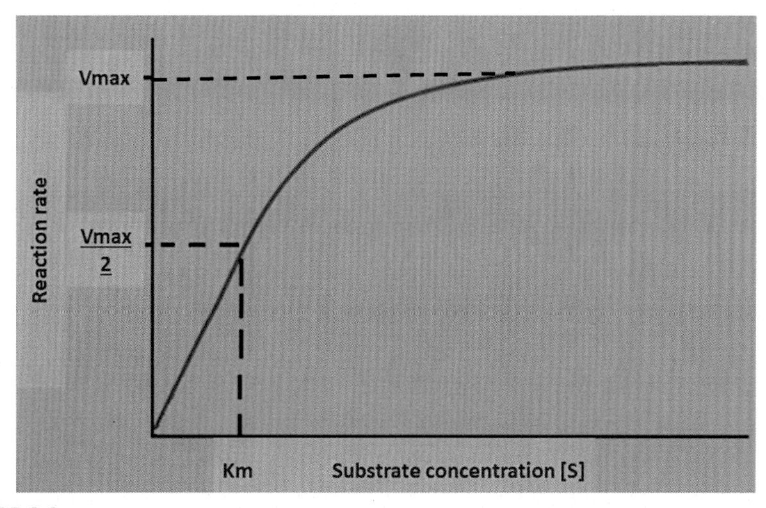

FIGURE 6.3

Relationship between initial substrate concentration and initial velocity.

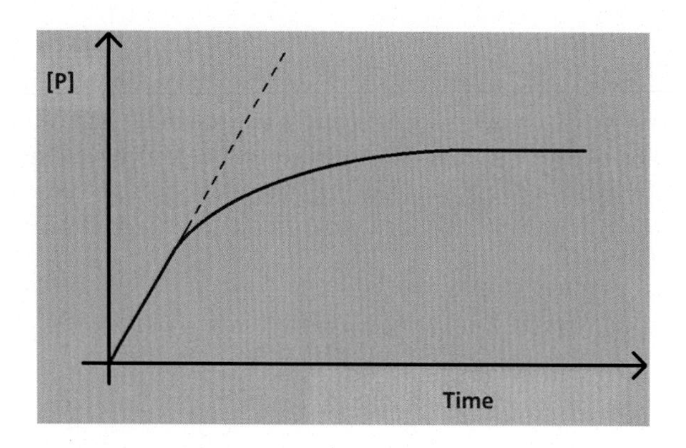

FIGURE 6.4

Relationship between product formation and velocity.

formation is analyzed by suitable measurable method with respect to time, the relationship can be represented graphically as given in Fig. 6.4.

The MM equation takes all the above features of enzyme catalyzed reaction as a base for framing its equation. Let us take an example of enzyme catalyzed reaction with single substrate.

$$E+S \underset{K_{-1}}{\overset{K_1}{\rightleftharpoons}} ES \underset{K_{-2}}{\overset{K_2}{\rightleftharpoons}} E+P \tag{6.1}$$

In the above equation, there are four independent reactions:

Reactions		Nature	Rate constants
Reaction 1	E +S ⟶ ES	Forward reaction	K_1
Reaction 2	ES ⟶ E + S	Backward reaction	K_{-1}
Reaction 3	ES ⟶ P	Forward reaction	K_2
Reaction 4	E + P ⟶ ES	Backward reaction	K_{-2}

The chances of reaction 4 to occur is negligible, therefore the equation becomes as follows:

$$E+S \underset{K_{-1}}{\overset{K_1}{\rightleftharpoons}} ES \xrightarrow{K_2} E+P \tag{6.2}$$

When the reaction just starts

$$\frac{d[ES]}{dt} = k_1[E][S] \tag{6.3}$$

where [ES] is the concentration of enzyme substrate complex, [E] is the concentration of free enzyme, and [S] is the concentration of substrate.

$$\frac{-d[ES]}{dt} = k_{-1}[ES] \tag{6.4}$$

According to reaction 2 given in the table, the rate of breakdown of ES can be given in form of the following.

For further calculations, we must remember that

$$[E_0] = [E] + [ES] \quad (6.5)$$

At initial stage-

There is equilibrium in stage one that is, enzyme, substrate and enzyme substrate formation (Product formation is slow).

The concentration of substrate [S] is much more than the enzyme concentration [E].

At the start point of the reaction, the product concentration [P] is nearly zero.

$$K_1[E][S] = K_{-1}[ES] \quad (6.6)$$

Taking constants to one side

$$\frac{[E][S]}{[ES]} = \frac{k_{-1}}{k_1} = ks \quad (6.7)$$

where K_S is dissociation constant of enzyme substrate complex.

$$[E_0] = [E] + [ES] \quad (6.8)$$

Total enzyme = Free enzyme + Bound enzyme

$$[E] = [E_o] - [ES] \quad (6.9)$$

Replacing Eq. (6.9) in (6.7)

$$\frac{([E_0] - [ES])[S]}{[ES]} = ks \quad (6.10)$$

$$ks[ES] = [E_0][S] - [ES][S] \quad (6.11)$$

$$ks[ES] + [ES][S] = [E_0][S] \quad (6.12)$$

$$ks[ES] + [ES][S] = [E_0][S] \quad (6.13)$$

$$[ES](ks + [S]) = [E_0][S] \quad (6.14)$$

$$ES = \frac{[E_0][S]}{[S] + ks} \quad (6.15)$$

$$V_0 = K_2[ES] \quad (6.16)$$

$$V_0 = K_2\frac{[E_0][S]}{[S] + ks} \quad (6.17)$$

V_{max} is the maximum velocity reached when enzyme is saturated with substrate that depends on

$$V_{max} = K_2[E_0] \quad (6.18)$$

Substituting the value of K_2 $[E_0]$ by V_{max}.

$$V_0 = \frac{V_{max}[S]_o}{[S]_o + ks} \tag{6.19}$$

The above is the most popular equation that explains the nature of enzyme-catalyzed reaction in detail. Although this equation forms a base, yet the enzyme catalyzed reaction do not follow the equilibrium of MM equation as they are very fast. The reciprocal of MM equation can be derived by plotting [S] with velocity of reaction. The MM equation graph does not provide accuracy for calculation of K_m and V_{max}. The reciprocal plot is called as LB plot. The LB plot was proposed in 1934 by Lineweaver and Burk as a reciprocal of MM plot.

They simply rectified the inaccuracy of MM by taking the reciprocals of all MM factors.

$$V_0 = \frac{V_{max}[S]_0}{[S]_0 + K_m} \tag{6.20}$$

$$\frac{1}{V_0} = \frac{[S]_0 + K_m}{V_{max}[S]_0} \tag{6.21}$$

$$\frac{1}{V_0} = \frac{[S]_0}{V_{max}[S]_0} + \frac{K_m}{V_{max}[S]_0} \tag{6.22}$$

$$\frac{1}{V_0} = \frac{1}{V_{max}} + \frac{K_m}{V_{max}} \frac{1}{[S]_0} \tag{6.23}$$

$$\frac{1}{V_0} = \frac{K_m}{V_{max}} g \frac{1}{[S]_0} + \frac{1}{V_{max}} \tag{6.24}$$

The MM equation and the reciprocal plot is given in Figs. 6.5 and 6.6. Fig. 6.6 explains the reciprocal plot in which K_m = Slope × V_{max}

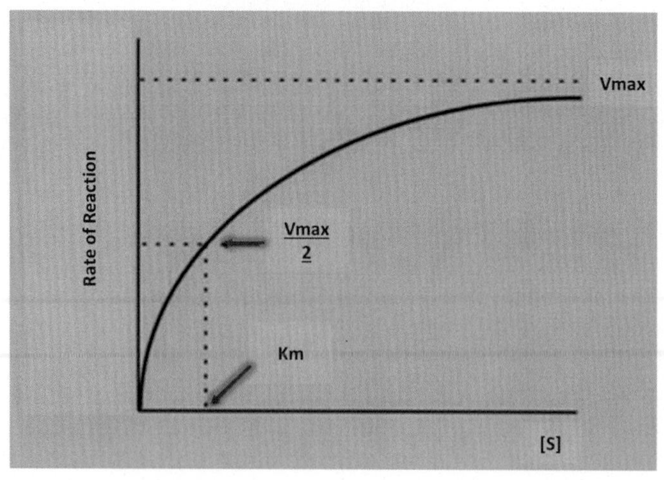

FIGURE 6.5

The MM equation.

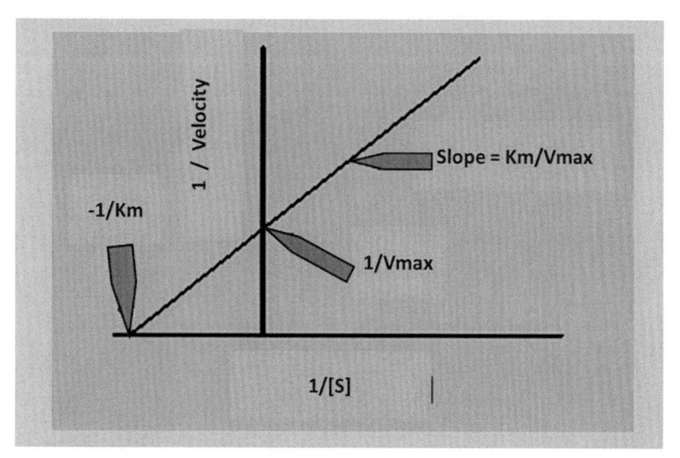

FIGURE 6.6

The reciprocal plot.

6.7 Advantages

A graph can be extrapolated without experiment being conducted till the substrate saturation.

If the graph is not linear, it indicates MM is not followed.

6.8 Disadvantages

$-1/K_m$ on X axis is sometimes difficult to plot.

The method relies more on the data at low substrate level concentration.

6.9 The significance of K_m and V_{max}

Metabolism analysis becomes handy by use of values of K_m and V_{max}. The purification studies and the kinetics of any enzyme involve determination of K_m and V_{max}. Following table enlists some enzymes and their related kinetic studies. Table 6.1 provides the detailed insight of fructosyltransferase, lipase, protease, and xylanase enzymes, their kinetic parameters and variance on accord of different sources.

Table 6.1 Kinetic parameters of some important industrial enzymes.

Source	K_m/V_{max}	Ref.
Fructosyltransferase		
Aspergillus phoenicis	Km 18 mM V_{max} 189 U/mg protein	Rustiguel et al. (2010)
Aspergillus aculeatus M105	K_m 2267 mM V_{max} 1347 μmol/min/mg K_{cat} 1550.2 /s,	Huang et al. (2016)
Pichia pastoris strain GS115	K_m 159.8 g/L V_{max} 0.66 g/(L_{min}),	Yang et al. (2016)
Aspergillus niger sp. *XOBP48*	K_m 79.51 mmol V_{max} 45.04 μmol/min K_{cat} 31.5/min	Ojwach et al. (2020)
Aspergillus niger ATCC 9642 and Penicillium brasilianum	K_m 24.60 mM V_{max} 104.16 μmol/min mL	Das Maso et al. (2020)
Aspergillus niger AS0023	K_m 44.38 mM V_{max} 1030 μmol/ mL/min	L'Hocine et al. (2000)
Rhodotorula sp. *LEB-V10*	MW 170 kDa	Hernalsteens and Maugeri (2008)
Lipase		
Geobacillus stearothermophilus AH22	*p*-Nitrophenyl esters K_m for (C2–C12) as substrates 0.16, 0.02, 0.19, and 0.55 mM V_{max} for 0.52, 1.03, 0.72, and 0.15 U mg	Ekinci et al. (2016)
Microbacterium sp.	K_m 3.2 mM V_{max} 50 μmol/min/mg	Tripathi et al. (2013)
Bacillus sp	K_m 0.31 mM V_{max} 7.6 μM/mL/min	Ghori et al. (2011)

Protease		
Bacillus mojavensis	K_m 0.0357 mg/mL (45°C) 0.0270 mg/mL (50°C) 0.0259 mg/mL (55°C) 0.0250 mg/mL (60°C V_{max} V_{max}: 74.07 99.01 116.28 120.48 μg/mL/min	Beg et al. (2002)
Aeribacillus pallidus C10	K_m 0.197 mg/mL V_{max} 7.29 l mol/mL/min	Yildirim et al. (2017)
Aspergillus sp	K_m Free and immobilized enzyme From 4.5 mg/mL to 5 mg/mL V_{max} From 200 U/mL to 370 U/mL	Sharma and Tripathi (2013)
Xylanase		
Bacillus subtilis- BS05	K_m 1.15 mg/mL K_{cat} 850 /s V_{max} 117.64 U/mg K_{cat}/K_m 739.13/ s/mg/mL	Irfan et al. (2013)
Streptomyces griseorubens LH-3	V_{max} 1.44 mg/mL K_m 2.05 μmol/min mg,	Wu et al. (2018)
Aspergillus oryzae HML366	K_m 1.16 mM V_{max} 336 μmol/min/mg	He et al. (2015)
Cladodes of Cereus pterogonus plant species	K_m 2.24 mg/mL V_{max} 5.8 μmol /min/mg.	Vikramathithan et al. (2010)
Streptomycetes strains	K_m 722 mol V_{max} 109 μmol/min	Shahid et al. (2012)
Bacillus velezensis AG20	V_{max} 21.0 ± 3.0 U/mL K_m 1.25 mg/mL K_{cat} 1.75/s	Amaro-Reyes et al. (2019)

6.10 Conclusion

The enzyme kinetics study is one of the critical steps required for evaluation of the enzyme performance and its industrial potential. Generally, the enzyme catalyzed reactions occur far from equilibrium. The MM equation provided the measure to evaluate the substrate affinity toward the enzyme as well as the maximum velocity of given enzyme-catalyzed reactions. The K_m should be used as an indicator to express steady state at given substrate concentration. Although linear part of the reaction is considered where the substrate concentration is too low. The complete enzyme kinetics depend on the initial stage of substrate transformation into product. The K_m is a very important value in biochemistry and enzymology to determine the rate of association of substrate that is of paramount importance in determining the enzyme performance.

Presently, alternative approaches are being constantly pursued for accuracy in judging the parameters in enzyme kinetics. Although the traditional enzyme kinetics methods are thought to be satisfying, yet due to enzyme engineering the variants that are generated need to be studied using suitable approaches for their satisfactory applicability. New methods have to be sought for accuracy to determine the enzyme kinetics parameters for proper categorization of the enzyme concerned.

References

Amaro-Reyes, A., Díaz-Hernández, A., Gracida, J., Blanca, E., Almendarez, G., García, M.E., Ochoa, T., Regalado, C., 2019. Enhanced performance of immobilized xylanase/filter paper-ase on a magnetic chitosan support. Catalysts 9 (11), 966.

Beg, Q.K., Saxena, R.K., Gupta, R., 2002. Kinetic constants determination for an alkaline protease from *Bacillus mojavensis* using response surface methodology. Biotechnol. Bioeng. 78 (3), 289–295.

Das Maso, S.S.S., Colet, R., Meirelles da Silva, L., Comin, T., Vendruscolo, M.D., Hassemer, G., Vargas, G.E.B., Lago, R.M.D., Backes, G.T., Zeni, J., 2020. Investigation of optimal conditions for production, characterization, and immobilization of fructosyltransferase and β-fructofuranosidase by filamentous fungi. Biointer. Res. Appl. Chem. 11 (4), 11187–11201.

Ekinci, A.P., Dinçer, B., Baltaş, N., Adıgüzel, A., 2016. Partial purification and characterization of lipase from Geobacillus stearothermophilus AH22. J. Enzym. Inhib. Med. Chem. 31 (2), 325–331. https://doi.org/10.3109/14756366.2015.1024677.

Ghori, M.I., Iqbal, M.J., Hameed, A., 2011. Characterization of a novel lipase from bacillus sp. Isolated from tannery wastes. Braz. J. Microbiol. 42, 22–29.

He, H., Qin, Y., Li, N., Chen, G., Liang, Z., 2015. Purification and characterization of a thermostable hypothetical xylanase from Aspergillus oryzae HML366. Appl. Biochem. Biotechnol. 175 (6), 3148–3161. https://doi.org/10.1007/s12010-014-1352-x.

Hernalsteens, S., Maugeri, F., 2008. Purification and characterisation of a fructosyltransferase from Rhodotorula sp. Appl. Microbiol. Biotechnol. 79 (4), 589–596. https://doi.org/10.1007/s00253-008-1470-x.

Huang, M.P., Wu, M., Xu, Q.S., Mo, D.J., Feng, J.X., 2016. Highly efficient synthesis of fructooligosaccharides by extracellular fructooligosaccharide-producing enzymes and immobilized cells of Aspergillus aculeatus M105 and purification and biochemical

characterization of a fructosyltransferase from the fungus. J. Agric. Food Chem. 64 (33), 6425–6432. https://doi.org/10.1021/acs.jafc.6b02115.
Irfan, M., Nadeem, M., Syed, Q., 2013. Purification and kinetics study of thermostable cellulase free xylanase from Bacillus subtilis. Protein Pept. Lett. 20 (11), 1225–1231. https://doi.org/10.2174/09298665113209990007.
Jogaiah, S., Praveen, S., De Britto, S., Konappa, N., Udayashankar, A.C., 2020. Exogenous priming of chitosan induces upregulation of phytohormones and resistance against cucumber powdery mildew disease is correlated with localized biosynthesis of defense enzymes. Int. J. Biol. Macromol. 162, 1825–1838. https://doi.org/10.1016/j.ijbiomac.2020.08.124.
L'Hocine, L., Wang, Z., Jiang, B., Xu, S., 2000. Purification and partial characterization of fructosyltransferase and invertase from Aspergillus Niger AS0023. J. Biotechnol. 81, 73–84. https://doi.org/10.1016/S0168-1656(00)00277-7.
Ojwach, J., Kumar, A., Mukaratirwa, S., Mutanda, T., 2020. Purification and biochemical characterization of an extracellular fructosyltransferase enzyme from *Aspergillus niger* sp. XOBP48: implication in fructooligosaccharide production. 3 Biotech 10 (10), 459. https://doi.org/10.1007/s13205-020-02440-w.
Rustiguel, C.B., Oliveira, A.H.C., Terenzi, H.F., Jorge, J.A., Guimaraes, L.H.S., 2010. Biochemical properties of an extracellular β-D-fructofuranosidase II produced by Aspergillus phoenicis under solid-sate fermentation using soy bran as substrate. Electron. J. Biotechnol. 14, 2. https://doi.org/10.2225/vol14-issue2-fulltext-1.
Shahid, G., Jabbar, A., Qureshi, M.Z., Aujla, M.I., Suleman, M., Sohail, M., 2012. Xylanase activity and kinetics comparison during pulping of Gossypium arboreum and Gossypium barbadense. Asian J. Chem. 24 (1), 441–443.
Sharma, N., Tripathi, N., 2013. Kinetic study of free and immobilized protease from Aspergillus sp. J. Pharm. Biol. Sci. 7 (2), 86–96.
Sudisha, J., Kumar, A., Amruthes, K.N., Niranjana, S.R., Shekar Shetty, H., 2011. Elicitation of resistance and defense related enzymes by raw cow milk and amino acids in pearl millet against downy mildew disease caused by *Sclerospora graminicola*. Crop Protect. 30, 794–801. https://doi.org/10.1016/j.cropro.2011.02.010.
Tripathi, R., Singha, J., Bhartia, R.K., Thakura, I.S., 2013. Isolation, purification and characterization of lipase from microbacterium sp. and its application in biodiesel production. 4th international conference on advances in energy research. Energy Proc. 54, 518–529.
Vikramathithan, J., Kumar, G.N., Muthuraman, P., Srikumar, K., 2010. Purification and characterization of thermophilic xylanase isolated from the xerophytic-Cereus pterogonus sp. Protein J. 29 (7), 481–486. https://doi.org/10.1007/s10930-010-9276-y.
Walsh, R., Martin, E., Darvesh, S., 2010. A method to describe enzyme-catalyzed reactions by combining steady state and time course enzyme kinetic parameters. Biochim. Biophys. Acta 1800, 1–5. https://doi.org/10.1016/j.bbagen.2009.10.007.
Wu, H., Cheng, X., Zhu, Y., Zeng, W., Chen, G., Liang, Z., 2018. Purification and characterization of a cellulase-free, thermostable endo-xylanase from Streptomyces griseorubens LH-3 and its use in biobleaching on eucalyptus kraft pulp. J. Biosci. Bioeng. 125 (1), 46–51. https://doi.org/10.1016/j.jbiosc.2017.08.006.
Yang, H., Wang, Y., Zhang, L., Shen, W., 2016. Heterologous expression and enzymatic characterization of fructosyltransferase from *Aspergillus niger* in *Pichia pastoris*. N. Biotech. 33, 164–170. https://doi.org/10.1016/j.nbt.2015.04.005.
Yildirim, V., Baltaci, M.O., Ozgencli, I., Sisecioglu, M., Adiguzel, A., Adiguzel, G., 2017. Purification and biochemical characterization of a novel thermostable serine alkaline protease from *Aeribacillus pallidus* C10: a potential additive for detergents. J. Enzym. Inhib. Med. Chem. 32, 468–477. https://doi.org/10.1080/14756366.2016.1261131.

CHAPTER 7

Protocols of important industrial enzymes

SUBCHAPTER 7.1

Large-scale production of enzyme—fructosyltransferase

Before you begin

Timing: For fungi—12 days.
For bacteria or yeast—96 h.

Media preparation

1. **Preparation of Czapeckdox agar plates**
   Dissolve sucrose 30.0 g, $NaNO_3$ 3.0 g, $MgSO_4$ 0.5 g, KCl 0.5 g, $FeSO_4$ 0.01 g, KH_2PO_4 1.0 g, distilled water 1000 mL, and adjust the pH to 5.50. Add 20 g agar-agar powder and homogenize.
   Sterilize the medium and prepare Czapeckdox agar plates for inoculation.
2. **Medium for spore suspension 0.95% (w/v) NaCl and 0.1% (v/v) Tween 80**
   For 100 mL of spore suspension medium, 0.95 g of NaCl and 0.11 mL (0.1 g) of Tween 80 are added to 100 mL of distilled water.
3. **Preparation of production medium**
   10 L of production medium, with the following composition (g L^{-1}): sucrose 320.500; yeast extract 2.107; urea 7.130; $MgSO_4.7H_2O$ 0.500; K_2HPO_4 5.000; $MnCl_2.4H_2O$ 0.030; $FeSO_4.7H_2O$ 0.010. 5 mL of polypropylene glycol were used as antifoam is prepared. Adjust the pH of the medium to 5.50.

Protocols and Applications in Enzymology. https://doi.org/10.1016/B978-0-323-91268-6.00009-0

Key resources table

Note that not all areas will be used in every protocol.

Reagent or resource	Source	Identifier
Biological samples		
Fungal culture		
Aspergillus niger	Isolated or	
Aspergillus oryzae	Stock culture	
Any fungus showing fructosyltransferase production in primary screening		

Materials and equipment

- **Materials:** Erlenmeyer flasks, volumetric flasks, conical flasks, funnel, test tubes, test tube stand, pipettes, Whatman filter paper 1, and pipette stand.
- **Equipments:** Batch bioreactor, weighing balance, autoclave, shaking incubator, shaking water bath, boiling water bath, and visible spectrophotometer
- **Alternatives**: Colorimeter can be used instead of spectrophotometers.

Step-by-step method details

Preparation of inoculum

Timing: Day 1—Day 7

1. Take the pure culture of the selected fungi.
2. Inoculate on the Czapeckdox plates.
3. After incubation for 7 days, scrap the spores and suspend it in a minimum amount of spore suspension medium.
4. After all the spores are transferred, dilute the suspension in adequate manner to receive 10^8 spores/mL.

   Note: (If bacteria is used for production then Day 1—Day 2).

Preparation of production medium

Timing: Day 7

1. The production medium is prepared and sterilized.

Inoculation of the production medium

Timing: Day 8

1. 10 L of production medium is taken in the bioreactor (take the medium in accordance to the capacity of the bioreactor you are working with).
2. After sterilization of the medium and the reactor, 100 mL of spore suspension is added and fermentation is initiated under strict growth conditions.

3. The temperature is set to 30°C (optimum temperature of the organism selected for production)
4. Agitation speed is adjusted to 800 rpm.
5. Fermentation period (4 days).
6. Adjust the aeration rate to 0.75 volume of air per volume of medium per minute (vvm).
7. Monitoring should be done by withdrawing samples after a fixed time interval (e.g., 12 h).
8. Filter the sample using Whatman Filter paper 1.
9. The filtrate is used as a crude enzyme source for assay.
10. The assay procedure is given in Chapter 5.
11. The total yield of the enzyme can be calculated once the fermentation is over.
12. A large amount of enzyme is produced, which can be further purified as per requirement and application

Pause point: All the incubation periods where the organism is given optimum conditions for growth and enzyme productions are the pause points.

Termination of fermentation

Timing: Day 12

Fermentation is stopped and final enzyme production is determined by assay.

Expected outcomes

Depending on the potential of the microbe selected for enzyme production, the enzyme is produced on large scale.

Quantification and statistical analysis

The standard protocol for assay will be followed as given in Chapter 5.

Advantages

The procedure is too simple.

Expertise is required for Bioreactor settings and operation.

Limitations

Success of the experiment depends on the potential of the fermenting microorganism chosen.

In the case of molds, the time required for enzyme production is prolonged due to a long incubation period.

Optimization and troubleshooting

Contamination is the most common problem encountered in fermentations that affect enzyme production.

Enzyme activity is not detected due to inhibitors produced by contaminants. Medium precipitates after sterilization, and no growth occurs. Medium is proper but no growth occurs. All these reasons hamper enzyme production, and no activity is registered in the crude sample.

Potential solution to optimize the procedure

If contamination occurs, there is no alternative then to discard and start a fresh.

Strict sterilization and aseptic transfers are the measures that can avoid contaminations.

Di-potassium hydrogen phosphate has to be added as a solution after all the ingredients are dissolved in a partial volume of water.

The medium should be properly cooled before inoculation.

Many times, the enzymes are sensitive to Fe salts. If all parameters are proper, then Fe salt can be avoided to cross-check to study its inhibitory effect on the enzyme.

Safety considerations and standards

Di-nitrosalicylic acid should be handled with care. It should not be allowed to come in contact with skin and eyes.

After the assay is complete, the fermentation remnants should be discarded after proper decontamination.

Alternative methods/procedures

Minor variations in the production media are witnessed.

Various bioreactors like airlift reactors and mechanically stirred reactors are experimented according to the need of the microorganism selected (Sedova et al., 2014).

The influence of aeration rate and agitation rate is investigated for optimization of enzyme yield (Maiorano et al., 2020).

References

Maiorano, A.E., da Silva, E.S., Perna, R.F., Ottoni, C.A., Piccoli, R.A.M., Fernandez, R.C., Maresma, B.G., Rodrigues, M.F.A., 2020. Effect of agitation speed and aeration rate on fructosyltransferase production of *Aspergillus oryzae* IPT-301 in stirred tank bioreactor. Biotechnol. Lett. 42, 2619–2629. https://doi.org/10.1007/s10529-020-03006-9.

Šedová, M., Illeová, V., Antošová, M., Annus, J., Polakovic, M., 2014. Production of fructosyltransferase in mechanically stirred and air-lift bioreactors. Chem. Pap. 68, 1639–1648. https://doi.org/10.2478/s11696-014-0563-5.

SUBCHAPTER

Assay of enzyme—xylanases 7.2

Before you begin

Timing: For fungi—12 days.
For bacteria or yeast—96 h.

Media preparation

4. **Preparation of Czapeckdox agar plates**
   Dissolve sucrose 30.0 g, $NaNO_3$ 3.0 g, $MgSO_4$ 0.5 g, KCl 0.5 g, $FeSO_4$ 0.01 g, KH_2PO_4 1.0 g, distilled water 1000 mL, and adjust the pH to 5.50. Add 20 g agar-agar powder and homogenize.
   Sterilize the medium and prepare Czapcckdox agar plates for inoculation.
5. **Medium for spore suspension 0.95% (w/v) NaCl and 0.1% (v/v) Tween 80**
   For 100 mL of spore suspension medium, 0.95 g of NaCl and 0.11 mL (0.1 g) of Tween 80 are added to 100 mL of distilled water.
6. **Preparation of the substrate for fermentation**
   Corncobs, wheat straw, or any other hemicellulose-rich agricultural waste can be used as a substrate for xylanase production. The selected substrate is washed dried at 60°C for 24 h. After complete drying, it is milled to a size of 1 cm (Diah et al., 2020). This step helps in the proper growth of microbes on the substrate surface as the area increases. The solid substrate was sterilized separately at 121°C for 15 min.
7. **Sodium acetate buffer (0.1M; pH 5.5):**
   a. **Solution A:** 1 M acetic acid was prepared by adding 12.06 mL of glacial acetic acid to 200 mL of distilled water.
   b. **Solution B:** 1 M NaOH was prepared by dissolving 4 g of NaOH in 100 mL of distilled water.

   146.0 mL of solution A and 100 mL of solution B were mixed together. The pH was adjusted with weak acid and alkali, and the final volume was made to 1000 mL with distilled water.
8. **Production medium (g/L):**
   Using the selected agroindustrial waste as carbon source 250 g, the other nutrients were of Prado medium $(NH_4)2SO_4$ 1.5 g/L; KH_2PO_4 2 g/L; urea 0.3 g/L; $CaCl_2$ 0.03 g/L; and $MgSO_4.7H_2O$ 0.2 g/L (Alves-Prado et al., 2010). The liquid nutrient supplement is sterilized before adding to the solid substrate. The pH of the medium should be adjusted to 6.0 by using 0.1 N HCl and 0.1 N NaOH.

Key resources table

Reagent or resource	Source	Identifier
Crude enzyme prepared in laboratory Commercial enzyme preparation of xylanase		
Fungal culture		
Aspergillus niger Aspergillus fumigatus Any fungus showing xylanase production in primary screening	Isolated or Stock culture	

Materials and equipment

- **Materials:** Volumetric flasks, conical flasks, funnel, test tubes, test tube stand, pipettes, and pipette stand.
- **Equipment:** Tray fermenter, weighing balance, autoclave, incubator, shaking water bath, boiling water bath, and visible spectrophotometer
- **Alternatives**: Colorimeter can be used instead of spectrophotometers.

Step-by-step method details

Preparation of inoculum

Timing: Day 1–Day 7

5. Take the pure culture of the selected fungi.
6. Inoculate on the Czapeckdox plates.
7. After incubation for 7 days, scrap the spores and suspend it in a minimum amount of spore suspension medium.
8. After all the spores are transferred, dilute the suspension in adequate manner to receive 10^8 spores/mL.

   Note: (If Bacteria is used for production then Day 1–Day 2).

Preparation of production medium

Timing: Day 7

1. The production medium is prepared and sterilized.

Inoculation of the production medium

Timing: Day 8

13. The substrate 250 g is placed on the tray (26 × 26 × 7 cm). The tray has 4 holes on the side for aeration. It is inoculated with the spore suspension 10^8/g of substrate. The liquid component is added to the solid substrate in the ratio of 4:1.
14. The trays are covered with cheesecloth with incubation at 30°C for 3 days.

Pause point: All the incubation periods where the organism is given optimum conditions for growth and enzyme productions are the pause points.

Termination of fermentation

Timing: Day 11

1. The fermentation is terminated, and the enzyme is extracted from the substrate bed.
2. The extraction is done by sodium acetate buffer pH 5.5 in the ratio of 1:8 (fermented substrate: Buffer) in a shaking incubator for 1 h.
3. Filter the sample with the cheesecloth.
4. Centrifuge at 4500 rpm for 30 min.
5. Use supernatant as enzyme source (Lakshmi et al., 2011).
6. The assay procedure is given in Chapter 5.
7. The final enzyme production is determined by assay.

Expected outcomes

Depending on the potential of the microbe selected for enzyme production, the enzyme is produced on small scale.

Quantification and statistical analysis

The standard protocol for assay will be followed as given in Chapter 5.

Advantages

The procedure is too simple.

No expertise is required.

Limitations

Success of the experiment depends on the potential of the fermenting microorganism chosen.

In the case of molds, the time required for enzyme production is prolonged due to a long incubation period.

Optimization and troubleshooting

This is the most common problem encountered in fermentations that affect enzyme production.

Enzyme activity is not detected due to inhibitors produced by contaminants. Medium precipitates after sterilization, and no growth occurs. Medium is proper but no growth occurs. All these reasons hamper enzyme production, and no activity is registered in the crude sample.

Potential solution to optimize the procedure

If contamination occurs, there is no alternative then to discard and start a fresh.

Strict sterilization and aseptic transfers are the measures that can avoid contaminations.

Di-potassium hydrogen phosphate has to be added as a solution after all the ingredients are dissolved in a partial volume of water.

The medium should be properly cooled before inoculation.

Many times the enzymes are sensitive to Fe salts. If all parameters are proper, then Fe salt can be avoided to cross-check to study its inhibitory effect on the enzyme.

Safety considerations and standards

Di-nitrosalicylic acid should be handled with care. It should not be allowed to come in contact with skin and eyes.

After the assay is complete, the fermentation remnants should be discarded after proper decontamination.

Alternative methods/procedures

Minor variations in the production media are witnessed.

For detailed knowledge about a variety of media that can be used for different organisms, the study of review is suggested (Motta et al., 2012).

References

Alves-Prado, H.F., Pavezzi, F.C., Leite, R.S.R., Oliveira, V.M., de Sette, L.D., De Silva, R., 2010. Screening and production study of microbial xylanase producers from Brazilian Cerrado. Appl. Biochem. Biotechnol. 161 (1–8), 333–346.

Diah, M., Penia, K., Tjandra, S., 2021. Evaluation of heat distribution and aeration on xylanase production from oil palm empty fruit bunches using tray bioreactor. IOP Conf. Ser.: Mater. Sci. Eng. 1143, 012013. https://doi.org/10.1088/1757-899X/1143/1/012013.

Lakshmi, G.S., Bhargavi, P., Shetty, R.P., 2011. Sustainable bioprocess evaluation for xylanase production by isolated Aspergillus terreus and Aspergillus fumigatus under solid—state fermentation using oil palm empty fruit bunch fiber. Curr. Trends Biotechnol. Pharm. 5, 1434—1444.

Motta, F.L., Andrade, C.C.P., Santana, M.H.A., 2012. A review of xylanase production by the fermentation of xylan. In: Classification, Characterization and Applications. https://doi.org/10.13140/RG.2.1.2781.6724.

Further reading

Meilany, D., Anugeraheni, D., Aziz, A., Kresnowati, M.T.A.P., Setiadi, T., 2020. The Effects of Operational Conditions in Scaling up of Xylanase Enzyme Production for Xylitol Production Reactor, vol. 20, pp. 32—37 (1).

SUBCHAPTER

Large-scale production of enzyme—lipase

7.3

Before you begin

Timing: For fungi—2 days.
For bacteria or yeast—96 h.

Media preparation

9. **Preparation of Czapeckdox agar plates**
   Dissolve sucrose 30.0 g, $NaNO_3$ 3.0 g, $MgSO_4$ 0.5 g, KCl 0.5 g, $FeSO_4$ 0.01 g, KH_2PO_4 1.0 g, distilled water 1000 mL, and adjust the pH to 5.50. Add 20 g agar-agar powder and homogenize.
   Sterilize the medium and prepare Czapeckdox agar plates for inoculation.
10. **Medium for spore suspension 0.95% (w/v) NaCl and 0.1% (v/v) Tween 80**
    For 100 mL of spore suspension medium, 0.95 g of NaCl and 0.11 mL (0.1 g) of Tween 80 are added to 100 mL of distilled water.
11. **Potassium phosphate buffer (0.1M; pH 6.0)**
    a. Take 800 mL of distilled water in a 1000 mL volumetric flask.
    b. Add 1.203 g of K_2HPO_4 to the water measured.
    c. Add 5.865 g of KH_2PO_4 to the above solution.
    d. Make up the volume to 1 L using distilled water.
12. **1% NaCl (w/v) and 1% Triton-X 100 (w/v) solution**
    Prepare 100 mL of the solution by adding 1 g NaCl and 1 g Triton-X in 100 mL of distilled water.
13. **Enzyme production medium (g/L)**
    A 1000 g of a suitable substrate (viz. rice husk, cottonseed cake, red gram, or mixture can be used) is taken in Tray Bioreactor (Nema et al., 2019). It is added with 0.05 M potassium phosphate buffer pH 6.0 to reach a moisture content of 75%.

Key resources table

Note that not all areas will be used in every protocol.

Reagent or resource	Source	Identifier
Biological samples		
Fungal culture		
Aspergillus niger	Isolated or Stock culture	
Any fungus showing fructosyltransferase production in primary screening		

Materials and equipment

- **Materials:** Erlenmeyer flasks, volumetric flasks, conical flasks, funnel, test tubes, test tube stand, Pipettes, Whatman filter paper 1, and pipette stand.
- **Equipments:** Tray bioreactor, weighing balance, autoclave, shaking incubator, shaking water bath, boiling water bath, and visible spectrophotometer
- **Alternatives**: Colorimeter can be used instead of spectrophotometers.

Step-by-step method details

Preparation of inoculum

Timing: Day 1–Day 7

9. Take the pure culture of the selected fungi.
10. Inoculate on the Czapeckdox plates.
11. After incubation for 7 days, scrap the spores and suspend it in a minimum amount of spore suspension medium.
12. After all the spores are transferred, dilute the suspension in adequate manner to receive 10^8 spores/mL.

Note: (If Bacteria is used for production then Day 1–Day 2).

Preparation of production medium

Timing: Day 7

1. The production medium is prepared and sterilized in trays and phosphate buffer pH 6.0 is added to maintain moisture 75%.

Inoculation of the production medium

Timing: Day 8

15. The autoclaved production medium is allowed to cool and inoculated with the 4% v/w of the above spore suspension as inoculum under aseptic conditions. The medium is incubated in laminar airflow at 30°C for 96 h.
16. The tray bioreactors are used for large-scale fermentations.
17. Place under the static condition at 30°C for 96 h (Edwinoliver et al., 2010).
18. Monitoring should be done by withdrawing 0.5 g of the sample after a fixed time 24 h interval (Doriya and Devarai, 2018).
19. Add 10 mL of 1% Triton X 100
20. Use Mortor and Pestle for grinding.
21. Filter the sample using Whatman Filter paper 1.
22. Take the filtrate and centrifuge at 10,000 rpm for 10 min.
23. The supernatant is used as a crude enzyme source for assay.
24. The assay procedure is given in Chapter 5.

Pause point: All the incubation periods where the organism is given optimum conditions for growth and enzyme productions are the pause points.

Termination of fermentation

Timing: Day 12

Fermentation is stopped and final enzyme production is determined by assay.

Expected outcomes

Depending on the potential of the microbe selected for enzyme production, the enzyme is produced on a large scale.

Quantification and statistical analysis

The standard protocol for assay will be followed as given in Chapter 5.

Advantages

The procedure is too simple.

No expertise is required.

Process of enzyme production is economic.

Limitations

Success of the experiment depends on the potential of the fermenting microorganism chosen.

In the case of molds, the time required for enzyme production is prolonged due to a long incubation period.

Optimization of the medium is required after small-scale production to set parameters such as temperature, pH, and inoculum size to maximize enzyme production.

Optimization and troubleshooting

Contamination is the most common problem encountered in fermentations that affect enzyme production.

Enzyme activity is not detected due to inhibitors produced by contaminants.

All these reasons hamper enzyme production, and no activity is registered in the crude sample.

Potential solution to optimize the procedure

If contamination occurs, there is no alternative then to discard and start a fresh.

Strict sterilization and aseptic transfers are the measures that can avoid contaminations.

Safety considerations and standards

After the assay is complete, the fermentation remnants should be discarded after proper decontamination.

Alternative methods/procedures

Minor variations in the production media are witnessed. Babassu cake, cane molasses, and many other raw materials are experimented for lipase production (Cavalcanti et al., 2005; Guttara, 2003).

References

Cavalcanti, Costa EAG., Estrada, M.L.F., Guimarães, D.M., Reis, L.D., Júnior, S.A., Lippel, G., 2005. Lipase production by solid-state fermentation in fixed-bed bioreactors. Braz. Arch. Biol. Technol. 48, 79–84. https://doi.org/10.1590/S1516-89132005000400010 spe.

Doriya, K., Devarai, S.K., 2018. Solid state fermentation of mixed substrate for L-asparaginase production using tray and in-house designed rotary bioreactor. Biochem. Eng. J. 138, 188–196.

Edwinoliver, N.G., Thirunavukarasu, K., Naidu, R.B., Gouthaman, M.K., Nakajima, K., Kamini, N.R., 2010. Scale up of a novel tri-substrate fermentation for enhanced production of *Aspergillus niger* lipase for tallow hydrolysis. Bioresour. Technol. 101, 6791–6796.

Gutarra, M.L.E., 2003. Produção de lipase por fermentação no estado sólido: seleção de fungos produtores e estudo das condições de cultivo. MSc Thesis. Departamento de Bioquímica, IQ/UFRJ, Rio de Janeiro/RJ, Brazil.

Nema, A., Patnala, S.H., Mandari, V., Kota, S., Devraj, S.K., 2019. Production and optimization of lipase using *Aspergillus niger* MTCC 872 by solid-state fermentation. Bull. Natl. Res. Cent. 43, 82. https://doi.org/10.1186/s42269-019-0125-7.

SUBCHAPTER

Large-scale production of protease 7.4

Before you begin

Timing: For bacteria 3 days.

Media preparation

14. **Preparation of production medium (mg/mL)**
    Prepare the production medium with composition glucose 4; Caesamino acid 8; KH_2PO_4 1; K_2HPO_4 3; Na_2SO_4 2; $MgSO_4.7H_2O$ 0.1. Add every component except gGlucose. Glucose is added at the end just before the fermentation begins (Beg et al., 2003).
15. **Medium for seed inoculum**
    2% of the total media is taken as inoculum. The medium is with the same composition. After inoculation of bacteria, incubate it and read absorbance at 550 nm (Nearly = 0.25). 24 h old culture is desirable for efficient enzyme production.

Key resources table

Reagent or resource	Source	Identifier
Crude enzyme prepared in laboratory Commercial enzyme preparation of xylanase		
Bacterial culture		
Bacillus subtilis	Isolated or Stock culture	
Any bacteria showing xylanase production in primary screening		

Materials and equipment

- **Materials:** Volumetric flasks, conical flasks, funnel, test tubes, test tube stand, pipettes, and pipette stand.
- **Equipments:** Stirred tank bioreactor, weighing balance, autoclave, incubator, shaking water bath, boiling water bath, and visible spectrophotometer
- **Alternatives**: Colorimeter can be used instead of spectrophotometers.

Step-by-step method details

Preparation of inoculum

Timing: Day 1

13. Take the pure culture of the selected bacteria.
14. Inoculate in the seed culture medium for 1 day.
15. Standardize the culture by taking absorbance at 550 nm.

Production of enzyme

Timing: Day 2

1. Prepare 10 L of production medium for 14 L of bioreactor.
2. Perform in situ sterilization at 121°C for 30 min.
3. Add the inoculum 2% (v/v) of the production medium taken in the bioreactor vessel.
4. Adjust the temperature at 50°C (or optimum temperature for your organism).
5. Set the fermentation with 24 h incubation period with sample withdrawals after every 2 h.
6. Agitation is 200 RPM, airflow 4 vvm, dissolved oxygen maintained at 20%.

Note: The parameters such as pH, temperature, agitation speed, and oxygen required differ from organism to organism. It is suggested that optimization of these parameters to be performed for the selected organism to produce maximum enzyme on large scale.

Termination and enzyme extraction

Timing: Day 3

25. The fermentation is terminated.
26. Centrifuge the broth at 10,000 rpm for 15 min.
27. The supernatant is the enzyme source.
28. The assay procedure is given in Chapter 5.

Pause point: As it is bacterial fermentation, all steps should be performed on each day.

Expected outcomes

Depending on the potential of the microbe selected for enzyme production, the enzyme is produced on large scale.

Quantification and statistical analysis

The standard protocol for assay will be followed as given in Chapter 5.

Advantages

The procedure is too simple.

No expertise is required.

As bacterial fermentation, it is completed in less time.

Enzyme production is very fast.

Limitations

Success of the experiment depends on the potential of the fermenting microorganism chosen.

Optimization and troubleshooting

Contamination is the most common problem encountered in fermentations that affect enzyme production.

Enzyme activity is not detected due to inhibitors produced by contaminants. Medium precipitates after sterilization, and no growth occurs. Medium is proper but no growth occurs. All these reasons hamper enzyme production, and no activity is registered in the crude sample.

Potential solution to optimize the procedure

If contamination occurs, there is no alternative then to discard and start a fresh.

Strict sterilization and aseptic transfers are the measures that can avoid contaminations.

The medium should be properly cooled before inoculation.

Safety considerations and standards

Reagents should be handled with care.

After the assay is complete, the fermentation residue should be discarded after proper decontamination.

Alternative methods/procedures

Minor variations in the production media are witnessed. The medium exemplified in this protocol uses simple components for enzyme production (Prasad et al., 2014).

Solid-state fermentations are also alternatives for enzyme production (Elumalai et al., 2020).

References

Beg, Q., Sahai, V., Gupta, R., 2003. Statistical media optimization and alkaline protease production from Bacillus mojavensis in bioreactor. Process Biochem. 39, 203–209. https://doi.org/10.1016/S0032-9592(03)00064-5.

Elumalai, P., Lim, J.M., Park, Y.J., Cho, M., Shea, P.J., Oh, B.-T., 2020. Agricultural waste materials enhance protease production by *Bacillus subtilis* B22 in submerged fermentation under blue light-emitting diodes. Bioprocess Biosyst. Eng. 43, 821–830. https://doi.org/10.1007/s00449-019-02277-5.

Prasad, R., Abraham, Koshy, T., Nair, Jayakumaran, A., 2014. Scale up of production in a bioreactor of a halotolerant protease from moderately halophilic Bacillus sp. isolated from soil. Braz. arch. biol. technol. Curitiba. 57, 448–455.

SUBCHAPTER

Small-scale production of enzyme—fructosyltransferase 7.5

Before you begin

Timing: X−X For fungi—12 days.
For bacteria or yeast—96 h.

Media preparation

16. **Preparation of Czapeckdox agar plates**
    Dissolve Sucrose 30.0 g, $NaNO_3$ 3.0 g, $MgSO_4$ 0.5 g, KCl 0.5 g, $FeSO_4$ 0.01 g, KH_2PO_4 1.0 g, distilled water 1000 mL, and adjust the pH to 5.50. Add 20 g agar-agar powder and homogenize.
    Sterilize the medium and prepare Czapeckdox agar plates for inoculation.
17. **Medium for spore suspension 0.95% (w/v) NaCl and 0.1% (v/v) Tween 80**
    For 100 mL of spore suspension medium, 0.95 g of NaCl and 0.11 mL (0.1g) of Tween 80 are added to 100 mL of distilled water.
18. **Enzyme production medium (g/L)**
    Dissolve sucrose 30.0 g, $NaNO_3$ 3.0 g, $MgSO_4$ 0.5 g, KCl 0.5 g, $FeSO_4$ 0.01 g, KH_2PO_4 1.0 g, distilled water 1000 mL, and adjust the pH to 5.50 for fungi and 7.0 for bacteria. Prepare 50 mL medium in 250 mL conical flask. All the ingredients except KH_2PO_4 should be dissolved in 500 mL of distilled water.
    KH_2PO_4 should be dissolved separately in a minimal amount of water and finally added to a solution containing other nutrients to avoid precipitation.
    The pH of the medium should be adjusted to 5.50 by using 0.1N HCl and 0.1N NaOH.
    The volume should be made up to 1 L with distilled water.
    The medium should be autoclaved at 10 $lbs/in.^2$ at 115°C for 20 min to avoid charring of carbohydrate content.

Key resources table

Note that not all areas will be used in every protocol.

Reagent or resource	Source	Identifier
Biological samples		
Fungal culture		
Aspergillus niger	Isolated or	
Trichoderma species	Stock culture	
Any fungus showing fructosyltransferase production in primary screening		

Materials and equipment

- **Materials:** Erlenmeyer flasks, volumetric flasks, conical flasks, funnel, test tubes, test tube stand, pipettes, Whatman filter paper 1, and Pipette stand.
- **Equipments:** Weighing balance, autoclave, shaking incubator, shaking water bath, boiling water bath, and visible spectrophotometer
- **Alternatives.** Colorimeter can be used instead of spectrophotometers

Step-by-step method details

Preparation of inoculum

Timing: Day 1–Day 7

16. Take the pure culture of the selected fungi.
17. Inoculate on the Czapeckdox plates.
18. After incubation for 7 days scrap the spores and suspend it in a minimum amount of spore suspension medium.
19. After all the spores are transferred, dilute the suspension in adequate manner to receive 10^8 spores/mL.

Note: (If Bacteria is used for production then Day1–Day 2).

Preparation of production medium

Timing: Day 7

1. The production medium is prepared and sterilized.

Inoculation of the production medium

Timing: Day 8

29. The autoclaved production medium is allowed to cool and inoculated with the 0.5 mL of the above spore suspension as inoculum under aseptic conditions. The medium is incubated at 27°C for 4 days.

30. The unbaffled Erlenmeyer flask serves as a small-scale fermentor.
31. Place in shaking incubator at 200 rpm at 30°C for 76 h.
32. Monitoring should be done by withdrawing samples after a fixed time interval (e.g., 12 h).
33. Filter the sample using Whatman Filter paper 1
34. The filtrate is used as a crude enzyme source for assay.
35. The assay procedure is given in Chapter 5.

Pause point: All the incubation periods where the organism is given optimum conditions for growth and enzyme productions are the pause points.

Termination of fermentation

Timing: Day 12

Fermentation is stopped and final enzyme production is determined by assay.

Expected outcomes

Depending on the potential of the microbe selected for enzyme production, the enzyme is produced on small scale.

Quantification and statistical analysis

The standard protocol for assay will be followed as given in Chapter 5.

Advantages

The procedure is too simple.

No expertise is required.

Limitations

Success of the experiment depends on the potential of the fermenting microorganism chosen.

In the case of molds, the time required for enzyme production is prolonged due to a long incubation period.

Optimization and troubleshooting

This is the most common problem encountered in fermentations that affect enzyme production.

Enzyme activity is not detected due to inhibitors produced by contaminants. Medium precipitates after sterilization, and no growth occurs. Medium is proper but no growth occurs. All these reasons hamper enzyme production, and no activity is registered in the crude sample.

Potential solution to optimize the procedure

If contamination occurs, there is no alternative then to discard and start a fresh.

Strict sterilization and aseptic transfers are the measures that can avoid contaminations.

Di-potassium hydrogen phosphate has to be added as a solution after all the ingredients are dissolved in a partial volume of water.

The medium should be properly cooled before inoculation.

Many times, the enzymes are sensitive to Fe salts. If all parameters are proper, then Fe salt can be avoided to cross-check to study its inhibitory effect on the enzyme.

Safety considerations and standards

Di nitro salicylic acid should be handled with care. It should not be allowed to come in contact with skin and eyes.

After the assay is complete, the fermentation remnants should be discarded after proper decontamination.

Alternative methods/procedures

Minor variations in the production media are witnessed.

Different researchers take different concentrations of sucrose (Lateef et al., 2012; Cunha et al., 2019). For detailed knowledge about a variety of media that can be used for different organisms, the study of review is suggested (Maiorano et al., 2008).

References

Cunha, J.S., Ottoni, C.A., Morales, Sergio A.V., Silva, Elda S., Maiorano, Alfredo E., Perna, Rafael F., 2019. Synthesis and characterization of fructosyltransferase from aspergillus oryzae ipt-301 for high fructooligosaccharides production. Braz. J. Chem. Eng. 36 (2), 657–668. https://doi.org/10.1590/0104-6632.20190362s20180572.

Lateef, A., Oloke, J., Gueguim-Kana, E.B., Raimi, O.R., 2012. Production of fructosyltransferase by a local isolate of *Aspergillus niger* in both submerged and solid substrate media. Acta Aliment. 40, 100–117. https://doi.org/10.1556/AAlim.41.2012.1.12.

Maiorano, A.E., Piccoli, R.M., da Silva, E.S., de Andrade Rodrigues, M.F., 2008. Microbial production of fructosyltransferases for synthesis of pre-biotics. Biotechnol. Lett. 30 (11), 1867–1877. https://doi.org/10.1007/s10529-008-9793-3.

SUBCHAPTER

Small-scale production of enzyme—lipase 7.6

Before you begin

Timing: For fungi—12 days.
For bacteria or yeast—96 h.

Media preparation

19. **Preparation of Czapeckdox agar plates**
    Dissolve sucrose 30.0 g, $NaNO_3$ 3.0 g, $MgSO_4$ 0.5 g, KCl 0.5 g, $FeSO_4$ 0.01 g, KH_2PO_4 1.0 g, distilled water 1000 mL, and adjust the pH to 5.50. Add 20 g agar-agar powder and homogenize.
    Sterilize the medium and prepare Czapeckdox agar plates for inoculation.
20. **Medium for spore suspension 0.95% (w/v) NaCl and 0.1% (v/v) Tween 80**
    For 100 mL of spore suspension medium, 0.95 g of NaCl and 0.11 mL (0.1 g) of Tween 80 are added to 100 mL of distilled water.
21. **Potassium phosphate buffer (0.1 M; pH 6.0)**
    a. Take 800 mL of distilled water in a 1000 mL volumetric flask.
    b. Add 1.203 g of K_2HPO_4 to the water measured.
    c. Add 5.865 g of KH_2PO_4 to the above solution.
    d. Make up the volume to 1 L using distilled water.
22. **1% NaCl (w/v) and 1% Triton-X 100 (w/v) solution**
    Prepare 100 mL of the solution by adding 1 g NaCl and 1 g Triton-X in 100 mL of distilled water.
23. **Enzyme production medium (g/L)**
    5 g of a suitable substrate (viz. Rice husk, Cotton seed cake, Red gram or mixture can be used) is taken in 250 mL of Erlenmeyer flask (Nema et al., 2019). It is added with 0.05 M potassium phosphate buffer pH 6.0 to reach a moisture content of 75%. Sterilize the flasks at 121°C for 15 min.

Key resources table

Note that not all areas will be used in every protocol.

Reagent or resource	Source	Identifier
Biological samples		
Fungal culture		
Aspergillus niger	Isolated or Stock culture	
Any fungus showing fructosyltransferase production in primary screening		

Materials and equipment

- **Materials:** Erlenmeyer flasks, volumetric flasks, conical flasks, funnel, test tubes, test tube stand, pipettes, Whatman filter paper 1 and pipette stand.
- **Equipments:** Weighing balance, autoclave, shaking incubator, shaking water bath, boiling water bath, and visible spectrophotometer.
- **Alternatives:** Colorimeter can be used instead of spectrophotometers.

Step-by-step method details

Preparation of inoculum

Timing: Day 1–Day 7

20. Take the pure culture of the selected fungi.
21. Inoculate on the Czapeckdox plates.
22. After incubation for 7 days, scrap the spores and suspend it in a minimum amount of spore suspension medium.
23. After all the spores are transferred, dilute the suspension in adequate manner to receive 10^8 spores/mL.

Note: (If bacteria is used for production then Day 1–Day 2).

Preparation of production medium

Timing: Day 7

1. The production medium is prepared and sterilized.

Inoculation of the production medium

Timing: Day 8

36. The autoclaved production medium is allowed to cool and inoculated with the 4% v/w of the above spore suspension as inoculum under aseptic conditions. The medium is incubated at 27°C for 96 h.
37. The unbaffled Erlenmeyer flask serves as a small-scale fermentor.
38. Place under the static condition at 30°C for 96 h (Edwinoliver et al., 2010).
39. Monitoring should be done by withdrawing 0.5 g of the sample after a fixed time 24 h interval (Doriya and Devarai, 2018).
40. Add 10 mL of 1% Triton X 100
41. Use Mortor and Pestle for grinding.
42. Filter the sample using Whatman Filter Paper 1.
43. Take the filtrate and centrifuge at 10,000 rpm for 10 min.
44. The supernatant is used as a crude enzyme source for assay.
45. The assay procedure is given in Chapter 5.

Pause point: All the incubation periods where the organism is given, optimum conditions for growth and enzyme productions are the pause points.

Termination of fermentation

Timing: Day 12

Fermentation is stopped, and final enzyme production is determined by assay.

Expected outcomes

Depending on the potential of the microbe selected for enzyme production, the enzyme is produced on small scale.

Quantification and statistical analysis

The standard protocol for assay will be followed as given in Chapter 5.

Advantages

The procedure is too simple.

No expertise is required.

Limitations

Success of the experiment depends on the potential of the fermenting microorganism chosen.

In the case of molds, the time required for enzyme production is prolonged due to a long incubation period.

Optimization of the medium is required if the enzyme production is too low or for further large-scale production.

Optimization and troubleshooting

This is the most common problem encountered in fermentations that affect enzyme production.

Enzyme activity is not detected due to inhibitors produced by contaminants. Medium precipitates after sterilization, and no growth occurs. Medium is proper but no growth occurs. All these reasons hamper enzyme production, and no activity is registered in the crude sample.

Potential solution to optimize the procedure

If contamination occurs, there is no alternative then to discard and start a fresh.

Strict sterilization and aseptic transfers are the measures that can avoid contaminations.

Safety considerations and standards

After the assay is complete, the fermentation remnants should be discarded after proper decontamination.

Alternative methods/procedures

Minor variations in the production media are witnessed. Babassu cake, cane molasses, and many other raw materials are experimented for lipase production (Elisa Cavalcanti et al., 2005; Guttara, 2003).

References

Cavalcanti, Elisa d'Avila Costa, Gutarra, Melissa Limoeiro Estrada, Freire, Denise Maria Guimarães, Castilho, Leda dos Reis, Sant'Anna Júnior, Geraldo Lippel, 2005. Lipase production by solid-state fermentation in fixed-bed bioreactors. Braz. Arch. Biol. Technol. 48, 79–84. https://doi.org/10.1590/S1516-89132005000400010 spe.

Doriya, K., Devarai, S.K., 2018. Solid state fermentation of mixed substrate for L-asparaginase production using tray and in-house designed rotary bioreactor. Biochem. Eng. J. 138, 188–196.

Edwinoliver, N.G., Thirunavukarasu, K., Naidu, R.B., Gouthaman, M.K., Nakajima, K., Kamini, N.R., 2010. Scale up of a novel tri-substrate fermentation for enhanced production of *Aspergillus niger* lipase for tallow hydrolysis. Bioresour. Technol. 101, 6791–6796.

Gutarra, M.L.E., 2003. Produção de lipase por fermentação no estado sólido: seleção de fungos produtores e estudo das condições de cultivo. MSc Thesis. Departamento de Bioquímica, IQ/UFRJ, Rio de Janeiro/RJ, Brazil.

Nema, A., Patnala, S.H., Mandari, V., Kota, S., Devaraj, S.K., 2019. Production and optimization of lipase using *Aspergillus niger* MTCC 872 by solid-state fermentation. Bull. Natl. Res. Cent. 43, 82. https://doi.org/10.1186/s42269-019-0125-7.

SUBCHAPTER

Small-scale production of protease 7.7

Before you begin

Timing: For bacteria 3 days.

Media preparation

24. Preparation of production medium
Prepare medium with casein 1%, yeast extract 0.5% and NaCl 2.5% and adjust the pH at 7.0. Take 50 mL of medium in 250 mL Erlenmeyer flask.

25. Medium for seed inoculum
Take 10 mL of casein 1%, yeast extract 0.5%, and NaCl 2.5 and prepare 1 day old culture of bacteria.

Key resources table

Reagent or resource	Source	Identifier
Crude enzyme prepared in laboratory Commercial enzyme preparation of xylanase		
Bacterial culture		
Bacillus subtilis Any bacteria showing xylanase production in primary screening	Isolated or Stock culture	

Materials and equipment

- **Materials:** Volumetric flasks, conical flasks, funnel, test tubes, test tube stand, pipettes, and pipette stand.
- **Equipments:** Weighing balance, autoclave, incubator, shaking water bath, boiling water bath, and visible spectrophotometer.
- **Alternatives**: Colorimeter can be used instead of spectrophotometers.

Step-by-step method details

Preparation of inoculum

Timing: Day 1

24. Take the pure culture of the selected bacteria.
25. Inoculate in the seed culture medium for 1 day.

Preparation of production medium and inoculation

Timing: Day 2

7. The production medium is prepared and sterilized.
8. Inoculum added is 1% of the production medium taken in the Erlenmeyer flask.
9. The Erlenmeyer flask is incubated at 30°C for 24 h under static conditions.

Termination and enzyme extraction

Timing: Day 3

46. The fermentation is terminated.
47. Centrifuge the broth at 10,000 rpm for 15 min.
48. The supernatant is the enzyme source.
49. The assay procedure is given in Chapter 5.

Pause point: As it is bacterial fermentation, all steps should be performed on each day.

Expected outcomes

Depending on the potential of the microbe selected for enzyme production, the enzyme is produced on small scale.

Quantification and statistical analysis

The standard protocol for assay will be followed as given in Chapter 5.

Advantages

The procedure is too simple.

No expertise is required.

As bacterial fermentation, it is completed in less time.

Enzyme production is very fast.

Limitations

Success of the experiment depends on the potential of the fermenting microorganism choosen.

Optimization and troubleshooting

Contamination is the most common problem encountered in fermentations that affect enzyme production.

Enzyme activity is not detected due to inhibitors produced by contaminants. Medium precipitates after sterilization, and no growth occurs. Medium is proper but no growth occurs. All these reasons hamper enzyme production, and no activity is registered in the crude sample.

Potential solution to optimize the procedure

If contamination occurs, there is no alternative then to discard and start a fresh.

Strict sterilization and aseptic transfers are the measures that can avoid contaminations.

The medium should be properly cooled before inoculation.

Safety considerations and standards

Reagents should be handled with care.

After the assay is complete, the fermentation residue should be discarded after proper decontamination.

Alternative methods/procedures

Minor variations in the production media are witnessed. The medium exemplified in this protocol uses simple components for enzyme production (Prasad et al., 2014).

Solid-state fermentations are also alternatives for enzyme production (Elumalai et al., 2020).

References

Elumalai, P., Lim, J.M., Park, Y.J., Cho, M., Shea, P.J., Oh, B.-T., 2020. Agricultural waste materials enhance protease production by *Bacillus subtilis* B22 in submerged fermentation under blue light-emitting diodes. Bioprocess Biosyst. Eng. 43, 821–830. https://doi.org/10.1007/s00449-019-02277-5.

Prasad, R., Abraham, Koshy, T., Nair, Jayakumaran, A., 2014. Scale up of production in a bioreactor of a halotolerant protease from moderately halophilic Bacillus sp. isolated from soil. Braz. Arch. Biol. Technol. 57, 448–455.

SUBCHAPTER

Assay of enzyme—xylanases 7.8

Before you begin

Timing: For fungi—12 days.
For bacteria or yeast—96 h.

Media preparation

26. Preparation of Czapeckdox agar plates
Dissolve sucrose 30.0 g, $NaNO_3$ 3.0 g, $MgSO_4$ 0.5 g, KCl 0.5g, $FeSO_4$ 0.01 g, KH_2PO_4 1.0 g, distilled water 1000 mL, and adjust the pH to 5.50. Add 20 g agar-agar powder and homogenize.
Sterilize the medium and prepare Czapeckdox agar plates for inoculation.

27. Medium for spore suspension 0.95% (w/v) NaCl and 0.1% (v/v) Tween 80
For 100 mL of spore suspension medium, 0.95 g of NaCl and 0.11 mL (0.1 g) of Tween 80 are added to 100 mL of distilled water.

28. Production medium (g/L): Dissolve Sucrose 30.0 g, $NaNO_3$ 3.0 g, $MgSO_4$ 0.5 g, KCl 0.5 g, $FeSO_4$ 0.01 g, KH_2PO_4 1.0 g, Birchwood Xylan 5 g, distilled water 1000 mL, and adjust the pH to 5.50. All the ingredients except KH_2PO_4 should be dissolved in 500 mL of distilled water. KH_2PO_4 should be dissolved separately in a minimal amount of water and finally added to a solution containing other nutrients to avoid precipitation.
The pH of the medium should be adjusted to 5.50 by using 0.1 N HCl and 0.1 N NaOH.
The volume should be made up to 1 L with distilled water.
The medium should be autoclaved at 10 lbs/in.^{2} at 115°C for 20 min to avoid charring of carbohydrate content.

Key resources table

Reagent or resource	Source	Identifier
Crude enzyme prepared in laboratory Commercial enzyme preparation of xylanase		
Fungal culture		
Aspergillus niger	Isolated or Stock culture	
Any fungus showing xylanase production in primary screening		

Materials and equipment

- **Materials:** Volumetric flasks, conical flasks, funnel, test tubes, test tube stand, pipettes, and pipette stand.
- **Equipments:** Weighing balance, autoclave, incubator, shaking water bath, boiling water bath, and visible spectrophotometer.
- **Alternatives**: Colorimeter can be used instead of spectrophotometers.

Step-by-step method details

Preparation of inoculum

Timing: Day 1−Day 7

26. Take the pure culture of the selected fungi.
27. Inoculate on the Czapeckdox plates.
28. After incubation for 7 days, scrap the spores and suspend them in a minimum amount of spore suspension medium.
29. After all the spores are transferred, dilute the suspension in adequate manner to receive 10^8 spores/mL.

Note: (If bacteria is used for production then Day 1−Day 2).

Preparation of production medium

Timing: Day 7

1. The production medium is prepared and sterilized.

Inoculation of the production medium

Timing: Day 8

50. The autoclaved production medium is allowed to cool and inoculated with the 0.5 mL of the above spore suspension as inoculum under aseptic conditions. The medium is incubated at 27°C for 4 days.
51. The unbaffled Erlenmeyer flask serves as a small-scale fermentor.
52. Place in shaking incubator at 200 rpm at 30°C for 76 h.
53. Monitoring should be done by withdrawing samples after a fixed time interval (e.g., 12 h).
54. Filter the sample using Whatman Filter Paper 1.
55. The filtrate is used as a crude enzyme source for assay.
56. The assay procedure is given in Chapter 5.

Pause point: All the incubation periods where the organism is given optimum conditions for growth and enzyme productions are the pause points.

Termination of fermentation

Timing: Day 12

Fermentation is stopped, and final enzyme production is determined by assay.

Expected outcomes

Depending on the potential of the microbe selected for enzyme production, the enzyme is produced on small scale.

Quantification and statistical analysis

The standard protocol for assay will be followed as given in Chapter 5.

Advantages

The procedure is too simple.

No expertise is required.

Limitations

Success of the experiment depends on the potential of the fermenting microorganism chosen.

In the case of molds, the time required for enzyme production is prolonged due to a long incubation period.

Optimization and troubleshooting

This is the most common problem encountered in fermentations that affect enzyme production.

Enzyme activity is not detected due to inhibitors produced by contaminants. Medium precipitates after sterilization, and no growth occurs. Medium is proper but no growth occurs. All these reasons hamper enzyme production, and no activity is registered in the crude sample.

Potential solution to optimize the procedure

If contamination occurs, there is no alternative then to discard and start a fresh.

Strict sterilization and aseptic transfers are the measures that can avoid contaminations.

Di-potassium hydrogen phosphate has to be added as a solution after all the ingredients are dissolved in a partial volume of water.

The medium should be properly cooled before inoculation.

Many times, the enzymes are sensitive to Fe salts. If all parameters are proper, then Fe salt can be avoided to cross-check to study its inhibitory effect on the enzyme.

Safety considerations and standards

Di-nitrosalicylic acid should be handled with care. It should not be allowed to come in contact with skin and eyes.

After the assay is complete, the fermentation remnants should be discarded after proper decontamination.

Alternative methods/procedures

Minor variations in the production media are witnessed.

Different researchers take different substrates like sugarcane baggasse (Irfan et al., 2016). For detailed knowledge about a variety of media that can be used for different organisms, the study of review is suggested (Motta et al., 2012).

References

Irfan, M., Asghar, U., Nadeem, M., Nelofer, R., Syed, Q., 2016. Optimization of process parameters for xylanase production by Bacillus sp. in submerged fermentation. J. Rad. Res. Appl. Sci. 9, 139–147.

Motta, F.L., Andrade, C.C.P., Santana, M.H.A., 2012. A review of xylanase production by the fermentation of xylan. In: Classification, Characterization and Applications. https://doi.org/10.13140/RG.2.1.2781.6724.

CHAPTER

Strategies to improve enzyme activity for industrial processes

8

8.1 Introduction

The industrial catalytical processes are passively getting transformed into biocatalytic types with increasing dependency on enzymes as a transformation tool. Invariably, in almost all sectors, enzymes are performing satisfactorily and better than their chemical transforming counterparts.

The important industries that have switched their processes to enzyme based are the food and beverage industry, baking industry, beverage industry, dairy industry, leather industry, and textile industry, which are the few to enlist. The fields of agriculture and medicine are also now diverting their processes to enzyme-based technologies as will be discussed in Chapters 11 and 12, respectively.

This intensive demand for catalytic tools is for a wide variety of enzymes with specific catalytic features befitting the industry type. Table 8.1: provides details of enzymes presently employed in various industrial sectors.

The vent of industrial requirement can be bridged by specific targeted enzyme production. The solution lies in the generation of an enzyme pool with the desired specific features. The novel type of enzymes can be resourced from the steps involved in microbial production of the enzyme viz innovations in the process parameters, screening methods, genetic resources, genetic modifications, and terminally enzyme engineering or protein engineering. Fig. 8.1: provides the milestone in enzyme production that can be experimented for enzyme quality enhancement.

8.1.1 Improvement of enzyme-based process

All natural processes tend to occur with ease in an aqueous environment. It was quite natural to conceptualize that enzymes also express their optimum catalytical potential in the water. As studied in Chapter 1, there is a role of the aqueous environment in folding and three-dimensional confirmation of the protein. The hydrophobic amino acids form an inner core that has all the amino acids participating in the active site gather together for catalysis. Initial experiments conducted on the replacement of normal catalytic medium with the nonaqueous medium was the fundamental step to decipher the unexplored properties of enzymes (Halling and Kvittingen, 1999; Halling, 2004; Carrea and Riva, 2000; Amorim and Peter, 2002; Kumar et al., 2016).

Protocols and Applications in Enzymology. https://doi.org/10.1016/B978-0-323-91268-6.00010-7

Table 8.1 Enzymes in industries.

Industry	Enzymes
Brewing	Amylases Proteases Cellulases Glucanases Pectinases
Dairy	Rennet Lipase Protease Peptidase
Juice industry	Catalase Glucose oxidase Acid protease Glucoamylase Polygalacturonase Pectin esterase Hemicellulase Carbohydrase
Meat processing industry	Acid protease Tyrosinase Glutaminase Elastase Papain Transglutaminase
Leather industry	Protease Carbohydrase Keratinase Lipase

The industrial demand for catalysis in nonaqueous environments was thrust that propelled research in this direction. The aqueous environment poses challenges in industrial catalysis at various levels such as insolubility of desired compounds, altered branch reactions, undesired byproducts, and unsatisfactory product recovery (Ke and Klibanov, 1998).

It was quite surprising that the absence of water exuberated enzyme action and stringent presence of an aqueous environment was not mandatory for catalysis (Grant, 2004). It was reported that the features of catalysis could be modulated by altering the environment of enzyme action. The enzyme displays maximum response toward the optimum pH and incorporation of activators in the entirely new catalytic environment. Initial experiments were conducted on specific enzymes but now experiments are widespread through especially hydrolases class of enzymes.

The nonaqueous environment was created by using hydrophobic solvents that enhance the enzyme activity as it does not totally dehydrate the enzyme and

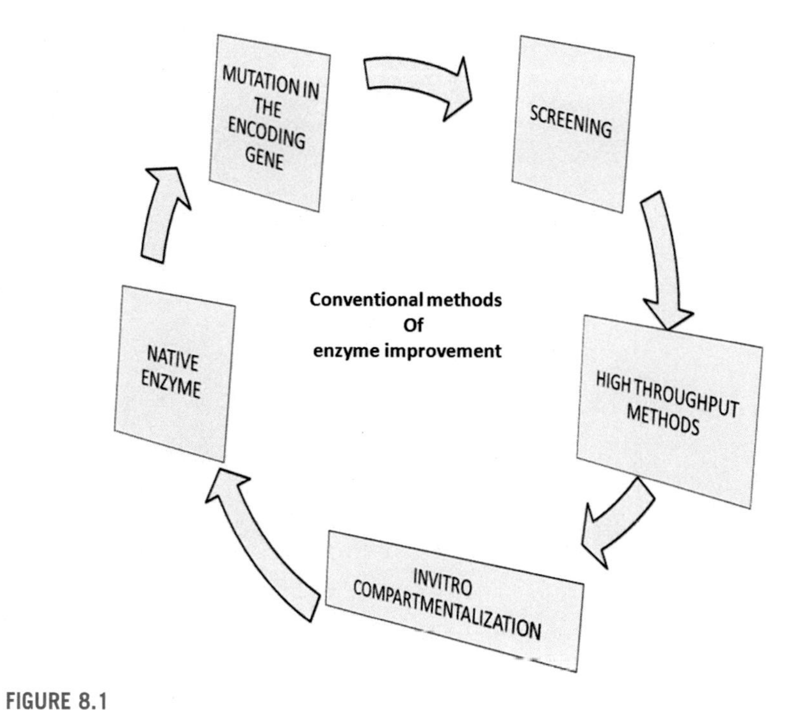

FIGURE 8.1

Strategies of enzyme enhancement.

dispenses water molecules critically important for activity when added to the system (Zaks and Klibanov, 1998a,b). The advanced strategies gaining popularity are breifed in Fig. 8.2.

Another important process modification is to provide ambient pH in the nonaqueous environment (Satpute et al., 2021). This is accomplished to freeze-dry the enzyme at desired pH accompanied by the protectant that helps the enzyme to rejuvenate in a nonaqueous environment with appropriate functionalities. Enzymes like lipases from sources such as *Bacillus*, *Candida rugosa*, and *Penicillium expansum* have been intensely experimented for optimization using process modification (Guncheva and Zhiryakova, 2011; Guncheva et al., 2011; Zhang et al., 2011; Murali et al., 2013).

Enzyme activity demands traces of water to execute the desired biotransformation. Excess of water in the aqueous environment does not play an efficient role and is probably not required for high enzyme activity (Jogaiah et al., 2011). Such enzymes that can exhibit activity in nearly nonaqueous medium have also been experimented (Clark Douglas, 2004). This microrequirement of water could be easily satisfied in organic solvents. According to Halling, it has an impact on the equilibrium of the reaction. The use of ionic liquids is also now attempted for enhancing enzyme activity. Ionic liquids are imparted with features like extremely low volatile nature, melting point below 100°C, noninflammable, and thermal stability as a virtue of its composition. These properties can be modulated by altering the composition

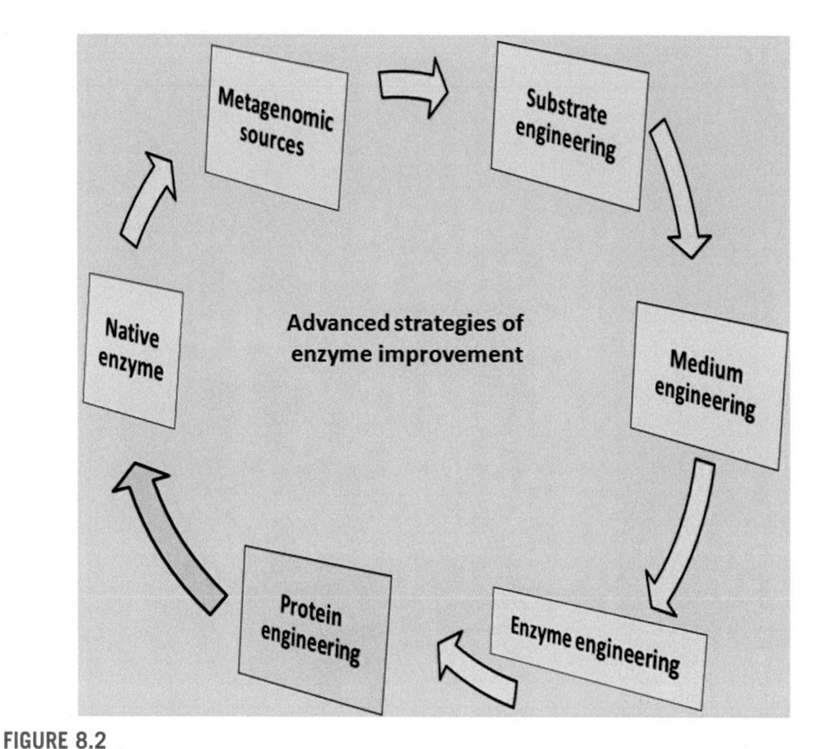

FIGURE 8.2

Advanced strategies of enzyme improvement.

and hence can be used as an effective tool. The enzyme coating by ionic liquids offers disadvantages like they are expensive and toxic (Fischer et al., 2011; Dominguez de Maria, 2008; Moniruzzaman et al., 2010).

8.1.2 Strategies to achieve efficient biocatalysis

8.1.2.1 Medium engineering

In 1987, Colja Laane focused on all possibilities of altering every factor that is related to enzyme action in the reaction medium. The term medium engineering was coined, which referred to the experimentation and influence of each and every medium component for enzyme action in a tailored format. It includes a maximum focus on all the microelements that on modulation will make the enzyme acquire maximum catalysis, stability, and product yield (Geetha et al., 2021). The amendment of the medium includes substrate, product flow, product yield, microenvironment requiring traces of water, forces of interaction, and their interplay between the substrate and the enzyme in the presence of the selected medium. This holistic approach of the medium constitution to achieve the target of maximum enzyme productivity is called as Medium Engineering (Laane, 1987). Fig. 8.3 explains the limitations offered by using an aqueous medium for the biotransformation of industrial compounds.

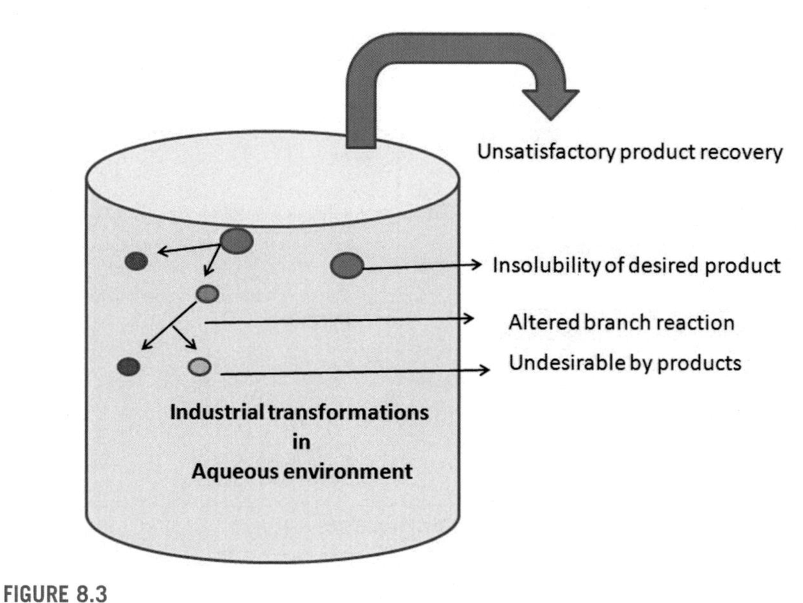

FIGURE 8.3

Aqueous medium for biotransformations.

Medium engineering has diverted toward the use of nonaqueous solvents as a premium approach for enzyme activity enhancement. The upcoming thrust areas of nonaqueous medium use are to extend the application of enzymes by overcoming barriers offered by an aqueous environment. The highlights of such medium replacements reported by various research groups are—extremely elevated level of product synthesis, improvised functionalities in the industrially prevalent environment, favorable reaction equilibrium, and inhibition of undesired aqueous phase side reactions above all stringent automated control of contamination. Fig. 8.4 explains the benefits of replacement of aqueous phase by nonaqueous phase.

There are two basic approaches: homogeneous biocatalysis and heterogeneous biocatalysis.

Homogeneous biocatalysis is a phenomenon where the enzyme exhibits respectable activity in a nonaqueous medium. The basic requirement of homogeneous biocatalysis in organic solvents is the solubility of enzymes in the selected solvent. Following are the main advantages and disadvantages of biocatalysis in organic solvents. Table 8.2 enlists the advantages and disadvantages of using homogeneous catalysis.

The nonaqueous medium presently employs enzymes in adsorbed form or cross-linked form or in powdered dry form. Three-dimensional structure and catalytic activity have been experimented in a wide range of enzymes using heterogeneous biocatalysis. Recently, studies are focused on the interactions of enzymes with their conjugate solvents (Rand, 2004; Gajbar et al., 2021). These interactions are tried to be engineered using polymer modification for enzyme Lipase A and polymer of

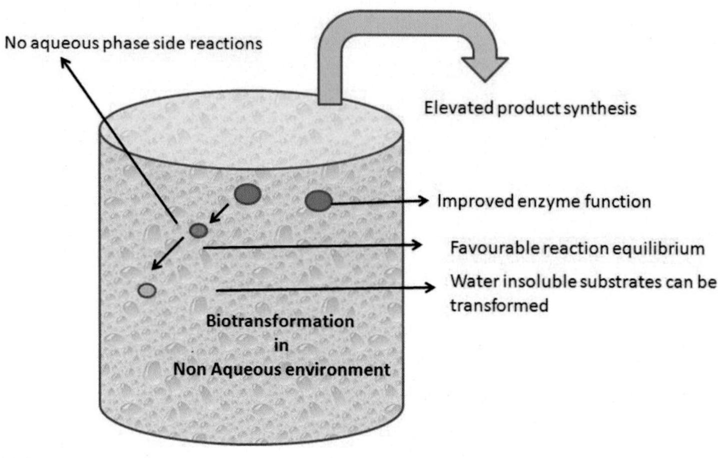

FIGURE 8.4

Advantages of nonaqueous medium for biotransformation.

Table 8.2 Homogeneous catalysis: advantages and disadvantages.

Advantages	Disadvantages
Excellent for reactions that could not be addressed in an aqueous environment	Recycling of the enzyme is a bit difficult
Change in the environment influences the direction of reaction in the case of a reversible reaction.	To enable recycling, covalent modification is desired
Influences the enantioselectivity	Modifications cause a hike in the cost
Influences substrate selectivity	In pure organic solvents, the enzyme exhibits restricted performance
Overcomes occurrence of aqueous side reactions	Maintenance of low water content Viscous solvents pose a problem in mass transfer
Enhanced activity in water organic mixtures	Involves more labor

acryloylmorpholine-co-*N*-isopropyl acrylamide. This modification of the enzyme aided in easy separation using reverse-phase temperature-based protocol (Chado et al., 2018). Table 8.3 provides a brief insight regarding enzymes successfully used with organic solvents.

The heterogeneous biocatalysis employs a carrier of the selected enzyme that in turn enables the enzyme-catalyzed reactions at more compromised activation energy levels than offered by the unbound enzyme (Adlercreutz, 1991). This further reduction in activation energy of the concerned reaction causes a high thrust in the enzyme productivity when compared to normal conditions. The alteration phase is selected

Table 8.3 Enzymes active in organic solvents.

Enzyme active in organic solvents	Ref
Bovine β-trypsin	Zhu et al. (2001)
Protease	Amorim and Peter (2002)
Horseradish peroxidase	Wei and Patricia (2002)
Subtilisin	Partridge et al. (1998)
Papain	Theppakorn et al. (2003)
Lipase	Valivety et al. (1992a,b)

such that it is different from the phase in which substrates and products exist. Fig. 8.5 details the features of heterogeneous catalysis.

Proper selection of the support for immobilization helps the enzyme to perform better, improves its stability in the reaction system, and provides the user an option of reusability of the enzyme (Govind et al., 2016). Approaches of solid surface adsorption is focussed on whole-cell adsorption and its biocatalysis. The latter is then engineered for improved kinetics and functional activity using principles of synthetic biology. The adsorption of enzymes on a surface involves congruent energy changes in comparison to those in protein foldings. The free energy is minimized by modulations in covalent bonding, hydrophobic interactions, ionic interactions, and weak forces of attraction that directs the levels of folding and three-dimensional structure of protein. The nonaqueous solvents and the dielectric constants also exerts profound influences on the protein conformation (Affleck et al., 1992). Fig. 8.5 provides features successfully used in heterogeneous biocatalysis.

The free energy changes, formation of bonds, and attachment of protein on solid surface involve multifactor influence. The utmost important factors are the surfaces that actually interact viz. property of the solid surface and the nature of the enzyme (Patel., 2014). Fig. 8.6 explains the nonaqueous biocatalysis and enzyme characteristics used for achieving it.

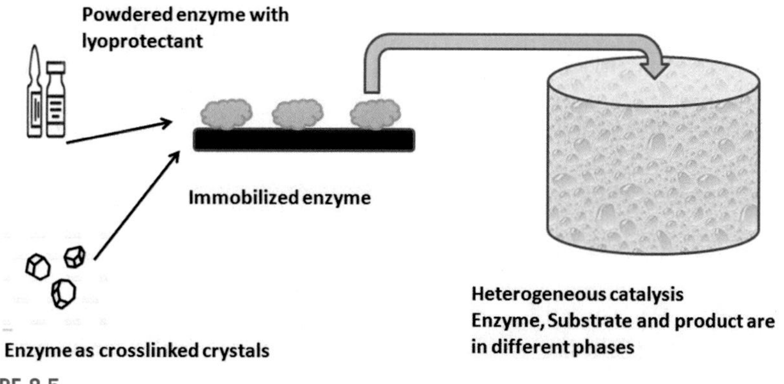

FIGURE 8.5

Features of enzymes used in heterogeneous biocatalysis.

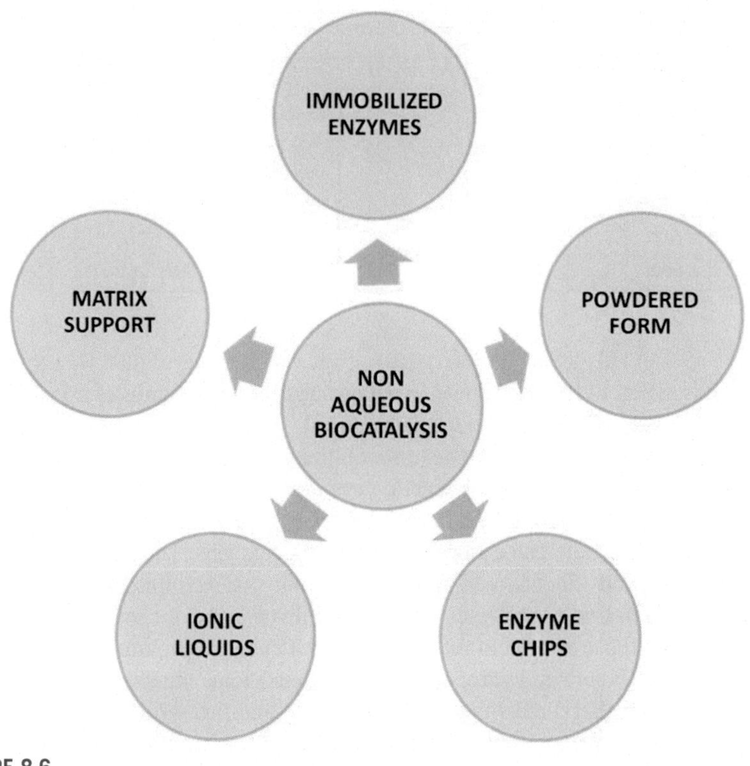

FIGURE 8.6

Enzyme forms used in nonaqueous biocatalysis.

8.1.3 Heterogeneous biocatalysis

Following are the important considerations in the development of heterogeneous catalysts

(a) Insight of principle of immobilization is important.
(b) In-depth knowledge about the structure and catalysis of the enzyme concerned is mandatory.
(c) The amount of enzyme successfully implanted on the solid support is a real engineering skill.
(d) Structural alteration after the enzyme transfer on a solid surface is very vital for future success.
(e) Characterization of the internal chemistry of the enzyme needs the intervention of technology.

Large-scale biotransformations are presently inviting heterogeneously catalyzed reactions in industries. All the process optimization is for the development of a continuous synthesis process with the capacity to reuse its catalytic agent that renders it an economical edge (Bolivar et al., 2016).

8.1.4 Substrate engineering

There are no perfect strategies developed yet which can deliver enzymes of the desired activity, stability, and above all specificity. Approaches to resolving this issue of deriving enhanced enzyme activity have led researchers to overcome the traditional methods having shortcomings by altering substrates now popularly termed as substrate engineering.

The substrate specificity of the enzymes sometimes offers hindrances in its applicability. Although, it is the strongest feature of the molecule that makes it an unparalleled catalyst. Engineering substrates have been an approach to achieve the maximization of production. If the functional group is attached to the desired substrate that confers its eligibility to get attached and acted upon by the selected enzyme, the expansion of substrate specificity is successful. In other words, newer compounds can be modified and brought under the umbrella of substrate for the selected efficient biotransformation molecule. The synthetic chemistry approach altered a compound by introducing a functional group. This modified compound was now acted upon by glycosyltransferase by just a modification initiated by binding eligibility. This method has the greatest advantages of increasing the inclusiveness of new compounds as substrates without disturbing the native protein form of the enzyme (Lairson et al., 2006).

8.1.5 Advanced strategies for enzyme improvement

The advanced strategies for improvement of enzyme production are essential to be experimented as the traditional approaches like mutations, isolation, reporter-based screening and selection (van Rossum et al., 2013), and improvisation in the fermentation process are incapable of rendering an industrially significant enzyme form. There are numerable setbacks associated with the conventional methods like the isolated strain is not stable or less potent. The labor required for mutation and selection of the strain is very laborious and time consuming.

The approach involving genetic engineering is attempted by various researchers for optimization of enzyme production by modulation at the genetic level. Cloning of the desired industrially important gene in the heterologous host is one of the promising options (Chen et al., 2007; Chen et al., 2007; Clementi and Rossi, 1986; Ding et al., 2002; He et al., 2009; Liu et al., 2005; Minning et al., 2001; Rodriguez et al., 2000a,b; Thongekkaew et al., 2008; Wonganu et al., 2008; Jogaiah et al., 2009). Pan et al. (2013) reported three novel approaches for improvising enzyme production.

(a) Introduction of strong promoter
Whenever a gene is introduced in a host, its expression is not so satisfactory due to a number of incompatibility grounds. Insertion of a strong promoter in the genetic makeup helps to improve the protein synthesis rate of the newly inserted gene. Glucoamylase has been improvised by inserting its gene to *Saccharomyces cerevisiae.* The use of plasmids with high copy number is also successfully approached for lipase production (Ashikari et al., 1989; Cha et al., 1992).

(b) Understanding the codons involved in enzyme expression
Codon bias is the term used to acknowledge the preference given by genetic expression to a particular codon during a particular gene expression. Such bais is possible to be executed due to degeneracy of codon. The host selected should be scrutinized for the execution of such codon bias for the desired enzymes. Amylases and lipases have been experimentally optimized in the selected heterologous host by using codon biasedness (Wang et al., 2001; Feng et al., 2002).

8.1.6 Altering the signal peptide

The signal peptide sequences also have exhibited to play role in the level of gene expression. Signal peptides have hydrophobic amino acids in the central region. Modifications of leader peptides were experimented as advanced strategies to enhance enzyme expression. Glucoamylase and phytase have been experimented for supporting the establishment of the role of signal peptide modifications on enzyme production (Liu et al., 2005; von Heijne, 1985, Xiong et al., 2003).

Bioinformatics has provided a platform for easy comparisons and conclusions of genetic sequences derived from new isolates with the existing database (Hallin et al., 2008). There are high throughput screening methods that help in screening the desired library in a limited time to come up with the desired mutant (Pushpam et al., 2011). The biocatalytic molecules with desirable properties can be hit with more possibilities using screening methods.

8.1.7 Metagenomics

One of the modern approaches is to explore the genetic source from microbes that fail to get cultivated in the laboratory (Sleator et al., 2008; Taku and Kentaro, 2009). Metagenomes from uncultured microorganisms are rich resources for novel enzyme encoding genes (Turnbaugh and Gordon, 2008). The methods used to screen the metagenomic libraries fall into two categories, which are based on the sequence or function of the enzymes. The sequence-based approaches rely on the known sequences of the target gene families (Simon and Daniel, 2011). In contrast, the function-based approaches do not involve the databases. There are basically two methods for screening of libraries and genomics: the first method is either to sequence the enzyme and the second method is based on the function of the enzyme. The function-based methods of deriving normal enzymes are discussed in the previous chapter that mainly consists of different screening strategies for identification of enzymes like the agar plate screening method, microtiter plates screening, ultrahigh throughput methods, in vitro compartmentalization with metagenomics, in vivo reporter-based screening methods (Leemhuis et al., 2009; Ferrer et al., 2009). Metagenomics can use genes that cannot be expressed in the laboratory as they are carried by the uncultivable (Kennedy et al., 2008; Jogaiah et al., 2013). Such sequences of enzymes can have any great industrial application.

The only limitation to use such metagenomic sequences is that they cannot be cultivated by normal methods used in laboratories (Singh et al., 2009; Vartoukian et al., 2010). There is a diverse range of explorable biological reservoirs that can code for enzymes that may have an absolute application in the industrial area. Fig. 8.7 enlists some specific enzymes used in the industrial sector.

Metagenomics has techniques that skip the culturing of microorganism and directly depend upon deriving the DNA samples from the environment (Handelsman, 2004; Simon and Daniel, 2009). Since its introduction, the use of metagenomics is improving the enzyme performance or helps to derive a novel enzyme that can satisfy industrial expectations. It is better to go for the method that is based upon function-based screening. The sequence-based screening method has a limitation that it would compare the data with the existing databases that can skip a very important sequence for potential enzyme activity.

8.1.8 Genetic engineering

Above all approaches elucidated in the chapter are no stronger than the use of genetic engineering. It is one of the strongest tools for engineering tailor-made proteins. The genetic strategies help to express the recombinant protein in model

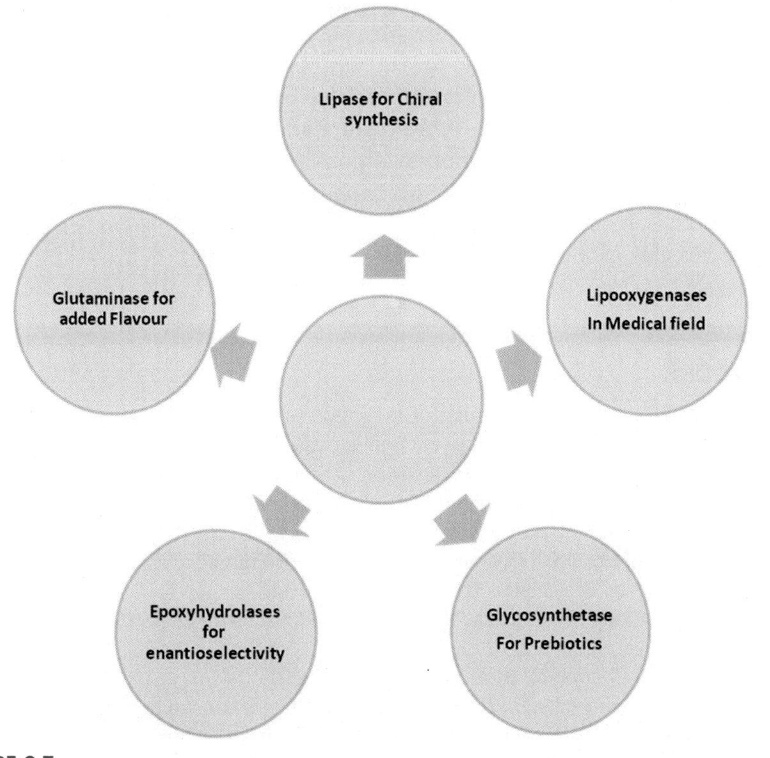

FIGURE 8.7

Industrially applied enzymes and their role.

organisms like *E. coli*. The recombinant protein is expressed in bulk as desired by industries (Sørensen and Mortensen, 2005). Many patents also have been registered for cloning and expressing the desired gene that would aid industrial processes viz phytase (Van Gorcom et al., 1995).

8.2 Conclusion

The biological catalysis has its premium time now as it is gaining wider acceptance in the market as well as at the user's end. The decade-old perception of enzymes as fragile, sensitive, and expensive catalytical tools has to be changed. Although there are many gray areas still unexplored by research regarding enzymes yet it has proved to be an invaluable molecular wealth, making life simpler and closer to environmental phenomena rather than chemical-based processes.

The most upcoming demand is nowadays for chiral compounds. The use of a nonaqueous environment is now a revolutionary approach to gather the unrevealed enzyme characteristics and novel features that could prove to be of immense industrial potential.

The immense research input in the upliftment of enzymes to satisfy the industrial sector still requires technological thrust. Many synthetic processes still do not come under the jurisdiction to be eligible for successful biotransformation. The medium modification with enhanced enzyme performance can be used. The engineering of substrates is also a novel option for synthetic biology reactions to be considered for biotransformation. Modulation of conditions to make the native enzyme form more robust and stable is one of the best approaches toward enzyme upliftment. Much research is to be conducted in substrate engineering and medium engineering to make it feasible as a widespread process.

References

Adlercreutz, P., 1991. On the importance of the support material for enzymatic synthesis in organic media. Support effects at controlled water activity. Eur. J. Biochem. 199, 609–614.

Affleck, R., Haynes, C.A., Clark, D.S., 1992. Solvent dielectric effects on protein dynamics. Proc. Natl. Acad. Sci. U. S. A 89, 5167–5170.

Amorim, F.F., Peter, H.J., 2002. Operational stability of high initial activity protease catalysts in organic solvents. Biotechnol. Prog. 18 (6), 1455–1457.

Ashikari, T., Kiuchi-Goto, N., Tanaka, Y., Shibano, Y., Amachi, T., et al., 1989. High expression and efficient secretion of *Rhizopus oryzae* glucoamylase in the yeast *Saccharomyces cerevisiae*. Appl. Microbiol. Biotechnol. 30, 515–520.

Bolivar, J.M., Eisl, I., Nidetzky, B., 2016. Advanced characterization of immobilized enzymes as heterogeneous biocatalysts. Catal. Today 259, 66–80.

Carrea, G., Riva, S., 2000. Properties and synthetic applications of enzymes in organic solvents. Angew Chem. Int. Ed. 39, 2226–2254.

Cha, H., Yoo, Y., Ahn, J., Kang, H., 1992. Expression of glucoamylase gene using SUC2 promoter in *Saccharomyces cerevisiae*. Biotechnol. Lett. 14, 747–752.

Chado, G.R., Holland, E.N., Tice, A.K., Stoykovich, M.P., Kaar, J.L., 2018. Exploiting the benefits of homogeneous and heterogeneous biocatalysis: tuning the molecular interaction of enzymes with solvents via polymer modification. ACS Catal. 8, 11579–11588. https://doi.org/10.1021/acscatal.8b03779.

Chen, J., Zhang, Y.Q., Zhao, C.Q., Li, A.N., Zhou, Q.X., et al., 2007. Cloning of a gene encoding thermostable glucoamylase from *Chaetomium thermophilum* and its expression in *Pichia pastoris*. J. Appl. Microbiol. 103, 2277–2284.

Chen, X., Cao, Y., Ding, Y., Lu, W., Li, D., 2007. Cloning, functional expression and characterization of *Aspergillus sulphureus* beta-mannanase in *Pichia pastoris*. J. Biotechnol. 128, 452–461.

Clark Douglas, S., 2004. Characteristics of nearly dry enzymes in organic solvents: implications for biocatalysis in the absence of water. Philos. Trans. R. Soc. Lond. B Biol. Sci. 99–105.

Clementi, F., Rossi, J., 1986. Alpha-amylase and glucoamylase production by *Schwanniomyces castellii*. Antonie Van Leeuwenhoek 52, 343–352.

Ding, S.J., Ge, W., Buswell, J.A., 2002. Secretion, purification and characterization of a recombinant *Volvariella volvacea* endoglucanase expressed in the yeast *Pichia pastoris*. Enzym. Microb. Technol. 31, 621–626.

Dominguez de Maria, P., 2008. Angew. Chem. Int. Ed. 47, 6960––6968.

Feng, D.G., Liu, X., Li, X.G., Zhu, Z., 2002. The relationship between tRNA abundance and gene expression. J. Chin. Biotechnol. 22, 1–8.

Ferrer, M., Beloqui, A., Vieites, J.M., Guazzaroni, M.E., Berger, I., Aharoni, A., 2009. Interplay of metagenomics and in vitro compartmentalization. Microb. Biotechnol. 2, 31–39.

Fischer, F., Mutschler, J., Zufferey, D.J., 2011. Ind. Microbiol. Biotechnol. 38, 477–487.

Gajbar, T.D., Kamble, M., Adhikari, S., Konappa, N., Satapute, P., Jogaiah, S., 2021. Gamma-irradiated fenugreek extracts mediates resistance to rice blast disease through modulating histochemical and biochemical changes. Anal. Biochem. 618, 114121.

Geetha, N., Bhavya, G., Abhijith, P., Shekhar, R., Dayananda, K., Jogaiah, S., 2021. Insights into nanomycoremediation: secretomics and mycogenic biopolymer nanocomposites for heavy metal detoxification. J. Hazard Mater. 409, 124541.

Govind, S.R., Jogaiah, S., Abdelrahman, M., Shetty, H.S., Tran, L.-S.P., 2016. Exogenous trehalose treatment enhances the activities of defense-related enzymes and triggers resistance against downy mildew disease of pearl millet. Front. Plant Sci. 7, 1593.

Grant, W.D., 2004. Life at low water activity. Philos. Trans. R. Soc. Lond. B Biol. Sci. 359, 1249–1267.

Guncheva, M., Zhiryakova, D., 2011. Catalytic properties and potential applications of *Bacillus* lipases. J. Mol. Catal. B Enzym. 68, 1–21.

Guncheva, M., Tashev, E., Zhiryakova, D., Tosheva, T., Tzokova, N., 2011. Immobilization of lipase from *Candida rugosa* on novel phosphorous-containing polyurethanes: application in wax ester synthesis. Process Biochem. 46, 923–930.

Hallin, P.F., Binnewies, T.T., Ussery, D.W., 2008. The genome BLASTatlas a GeneWiz extension for visualization of whole-genome homology. Mol. Biosyst. 4, 363–371.

Halling, P.J., Kvittingen, L., 1999. Why did biocatalysis in organic media not take off in the 1930's? Trends Biotechnol. 17, 343–344.

Halling, P.J., 2004. What can we learn by studying enzymes in non-aqueous media? Philos. Trans. R. Soc. Lond. B Biol. Sci. 359, 1287–1328. https://doi.org/10.1098/rstb.2004.1505.

Handelsman, J., 2004. Metagenomics: application of genomics to uncultured microorganisms. Microbiol. Mol. Biol. Rev. 68, 669–685.

He, J., Yu, B., Zhang, K., Ding, X., Chen, D., 2009. Expression of endo-1,4-beta-xylanase from *Trichoderma reesei* in *Pichia pastoris* and functional characterization of the produced enzyme. BMC Biotechnol. 9, 56.

Jogaiah, S., Ananda Kumar, S., Shetty, N.P., Shekar Shetty, H., 2009. Cloning and development of pathotype specific SCAR marker associated to *Sclerospora graminicola* isolates from pearl millet. Australas. Plant Pathol. 38, 216–221.

Jogaiah, S., Kumar, A., Amruthesh, K.N., Niranjana, S.R., Shekar Shetty, H., 2011. Elicitation of resistance and defense related enzymes by raw cow milk and amino acids in pearl millet against downy mildew disease caused by *Sclerospora graminicola*. Crop Protect. 30, 794–801.

Jogaiah, S., Sharathchandra, R.G., Lam-Son, P.T., 2013. Systems biology-based approaches towards understanding drought tolerance in food crops. Crit. Rev. Biotechnol. 33, 23–39.

Ke, T., Klibanov, A.M., 1998. On enzymatic activity in organic solvents as a function of enzyme history. Biotechnol. Bioeng. 57, 746–750.

Kennedy, J., Marchesi, J.R., Dobson, A.D., 2008. Marine metagenomics: strategies for the discovery of novel enzymes with biotechnological applications from marine environments. Microb. Cell Factories 7, 27.

Kumar, A., Dhar, K., Kanwar, S.S., Arora, P.K., 2016. Lipase catalysis in organic solvents: advantages and applications. Biol. Proced. Online 18, 2. https://doi.org/10.1186/s12575-016-0033-2.

Laane, C., 1987. Medium-engineering for bio-organic synthesis. Biocatalysis 1, 17–22. https://doi.org/10.3109/10242428709040127.

Lairson, L.L., Watts, A.G., Wakarchuk, W.W., Withers, S.G., 2006. Using substrate engineering to harness enzymatic promiscuity and expand biological catalysis. Nat. Chem. Biol. 2, 724–728. https://doi.org/10.1038/nchembio828.

Leemhuis, H., Kelly, R.M., Dijkhuizen, L., 2009. Directed evolution of enzymes: library screening strategies. IUBMB Life 61, 222–228.

Liu, S.H., Chou, W.I., Sheu, C.C., Chang, M.D., 2005. Improved secretory production of glucoamylase in Pichia pastoris by combination of genetic manipulations. Biochem. Biophys. Res. Commun. 326, 817–824.

Minning, S., Serrano, A., Ferrer, P., Solá, C., Schmid, R.D., et al., 2001. Optimization of the high-level production of Rhizopus oryzae lipase in Pichia pastoris. J. Biotechnol. 86, 59–70.

Moniruzzaman, M., Nakashima, K., Kamiya, N., Goto, M., 2010. Recent advances of enzymatic reactions in inonic liquids. Biochem. Eng. J. 48, 295–314.

Murali, M., Jogaiah, S., Amruthesh, K.N., Shinichi, I., Shekar Shetty, H., 2013. Rhizosphere fungus *Penicillium chrysogenum* promotes growth and induces defense-related genes and downy mildew disease resistance in pearl millet. Plant Biol. 15, 111–118.

Pan, H., Chen, Y., Yu, P., 2013. Advanced strategies for improving the production of industrial enzymes in heterologous host systems. Enzyme Eng. 2, 114. https://doi.org/10.4172/2329-6674.1000114.

Partridge, J., Dennison, P.R., Moore, B.D., Halling, P.J., 1998. Activity and mobility of subtilisin in low water organic media: hydration is more important than solvent dielectric. Biochim. Biophys. Acta 1386, 79–89.

Patel, N., 2014. Engineering Heterogeneous Biocatalysis Chemical Engineering. Ph.D. thesis. Columbia University, USA.

Pushpam, P.L., Rajesh, T., Gunasekaran, P., 2011. Identification and characterization of alkaline serine protease from goat skin surface metagenome. Amb. Express 1, 3.

Rand, R.P., 2004. Probing the role of water in protein conformation and function. Philos. Trans. R. Soc. Lond. B Biol. Sci. 359, 1277–1285.

Rodriguez, E., Wood, Z.A., Karplus, P.A., Lei, X.G., 2000a. Site-directed mutagenesis improves catalytic efficiency and thermostability of *Escherichia coli* pH 2.5 acid phosphatase/phytase expressed in *Pichia pastoris*. Arch. Biochem. Biophys. 382, 105–112.

Rodriguez, E., Mullaney, E.J., Lei, X.G., 2000b. Expression of the *Aspergillus fumigatus* phytase gene in *Pichia pastoris* and characterization of there combinant enzyme. Biochem. Biophys. Res. Commun. 268, 373–378.

Satpute, D.P., Vaidya, G.N., Lokhande, S.K., Shinde, S.D., Bhujbal, S.M., Chatterjee, D.R., et al., 2021. Organic reactions in aqueous media catalyzed by nickel. Green Chem. 23, 6273–6300.

Simon, C., Daniel, R., 2009. Achievements and new knowledge unraveled by metagenomic approaches. Appl. Microbiol. Biotechnol. 85, 265–276.

Simon, C., Daniel, R., 2011. Metagenomic analyses: past and future trends. Appl. Environ. Microbiol. 77, 1153–1161.

Singh, J., Behal, A., Singla, N., Joshi, A., Birbian, N., Singh, S., et al., 2009. Metagenomics: concept, methodology, ecological inference and recent advances. Biotechnol. J. 4, 480–494.

Sleator, R.D., Shortall, C., Hill, C., 2008. Metagenomics. Lett. Appl. Microbiol. 47, 361–366.

Sørensen, H.P., Mortensen, K.K., 2005. Advanced genetic strategies for recombinant protein expression in *Escherichia coli*. J. Biotechnol. 115, 113–128.

Taku, U., Kentaro, M., 2009. Functional metagenomics for enzyme discovery: challenges to efficient screening. Curr. Opin. Biotechnol. 20 (6), 616–622.

Theppakorn, T., Kanasawud, P., Halling, P.J., 2003. Effect of solid-state buffers on the catalytic activity of papain in low-water media. Enzyme Microb. Technol. 32, 828–836.

Thongekkaew, J., Ikeda, H., Masaki, K., Iefuji, H., 2008. An acidic and thermostable carboxymethyl cellulase from the yeast Cryptococcus sp. S-2: purification, characterization and improvement of its recombinant enzyme production by high cell-density fermentation of *Pichia pastoris*. Protein Expr. Purif. 60, 140–146.

Turnbaugh, P.J., Gordon, J.I., 2008. An invitation to the marriage of metagenomics and metabolomics. Cell 134, 708–713.

Valivety, R.H., Halling, P.J., Macrae, A.R., 1992a. Rhizomucor miehei lipase remains highly active at water activity below 0.0001. FEBS Lett. 301, 258–260.

Valivety, R.H., Halling, P.J., Peilow, A.D., Macrae, A.R., 1992b. Lipases from different sources vary widely in dependence of catalytic activity on water activity. Biochim. Biophys. Acta 122, 143–146.

Van Gorcom, R.F.M., Van Hartingsveldt, W., Van Paridon, P.A., Veenstra, A.E., Luiten, R.G.M., Selten, G.C.M., 1995. Cloning and expression of phytase from Aspergillus. US Patent, 5436156.

van Rossum, T., Kengen, S.W., van der Oost, J., 2013. Reporter-based screening and selection of enzymes. FEBS J. 280, 2979–2996.

Vartoukian, S.R., Palmer, R.M., Wade, W.G., 2010. Strategies for culture of 'unculturable' bacteria. FEMS Microbiol. Lett. 309, 1–7.

von Heijne, G., 1985. Signal sequences. The limits of variation. J. Mol. Biol. 184, 99–105.

Wang, H., Wu, Q., Liu, S., Xie, J., Ma, M., 2001. Cloning and sequence analysis of the phytase phyA gene of *Aspergillus niger* N25. Weishengwu Xuebao 41, 310–314.

Wei, G., Patricia, M.A., 2002. Raman evidence that the lyoprotectant poly(ethylene glycol) does not restore nativity to the heme active site of horseradish peroxidase suspended in organic solvents. Biomacromolecules 3, 846–849.

Wonganu, B., Pootanakit, K., Boonyapakron, K., Champreda, V., TanapongpipatS, et al., 2008. Cloning, expression and characterization of a thermotolerant endoglucanase from *Syncephalastrum racemosum* (BCC18080) in *Pichia pastoris*. Protein Express. Purif. 58, 78–86.

Xiong, A.S., Peng, R.H., Li, X., Fan, H.Q., Yao, Q.H., et al., 2003. Influence of signal peptide sequences on the expression of heterogeneous proteins in *Pichia pastoris*. Sheng Wu Hua Xue Yu Sheng Wu Wu Li Xue Bao 35, 154–160.

Zaks, A., Klibanov, A.M., 1998a. Enzymatic catalysis in nonaqueous solvents. J. Biol. Chem. 263, 3194–3201.

Zaks, A., Klibanov, A.M., 1998b. The effect of water on enzyme action in organic media. J. Biol. Chem. 263, 8017–8021.

Zhang, K.P., Lai, J.Q., Huang, Z.L., Yang, Z., 2011. *Penicillium expansum* lipase-catalyzed production of biodiesel in ionic liquids. Bioresour. Technol. 102, 2767–2772.

Zhu, G., Huang, Q., Zhu, Y., Li, Y., Chi, C., Tang, Y., 2001. X-Ray study on an artificial mung bean inhibitor complex with bovine beta-trypsin in neat cyclohexane. Biochim. Biophys. Acta 1546, 98–106.

CHAPTER

Scope and relevance of industrial applications

9

9.1 Introduction

Since human civilization, catalysis (whole cell and cell-free) has been used to aid human life. The applications were limited to brewing, baking, and cheese-making processes. With the expansion of the knowledge especially after the cell-free fermentation was acknowledged, the widespread applicability of enzymes was conceptualized as of using whole cells for transformation or direct use of enzymes. Presently, advances in biotransformation depend on recombinant DNA technology that renders efficient biocatalyst molecules that are highly specific and could be synthesized in bulk. The successful large-scale production of enzymes has enabled their uses in biotransformation at the industrial level. The major industries deriving benefits are food, fodder, paper, pulp, leather, textiles, etc.

The trend in the enzyme market reveals that the maximum demand is of protease and lipase enzymes followed by amylase and cellulases. The use of enzymes as a resource helps to use waste materials as raw materials, saves energy, and confers economic viability to the transformation process with reduced consumption of water and energy. Above all, it helps the environment-friendly transformation process to occur at a large scale.

The global market categorizes the industrial enzyme sector on the basis of enzyme used or enzyme-involved products launched. On basis of enzymes, the market categories experiencing a boom are lipases, carbohydrases, polymerases, and nucleases. Depending on the enzyme applicability, the major upcoming industrial sectors are shown in Fig. 9.1.

According to the EMR report, the major industrial market for enzymes is based in North America, Asia Pacific, Africa, Europe, and the Middle East. The major companies dominating enzyme production and supply are Novozymes, DSM Nutritional Products, Amano Enzymes Corporation, Advanced Enzymes USA, AB Enzymes GmbH, MetGenOy, Aum Enzymes, and many more. According to the market forecast, the market size value in 2020 was $5.93 billion and is expected to reach $9.14 billion in 2027. The prediction of a rise in CAGR is 7.5% till 2026.

Protocols and Applications in Enzymology. https://doi.org/10.1016/B978-0-323-91268-6.00011-9

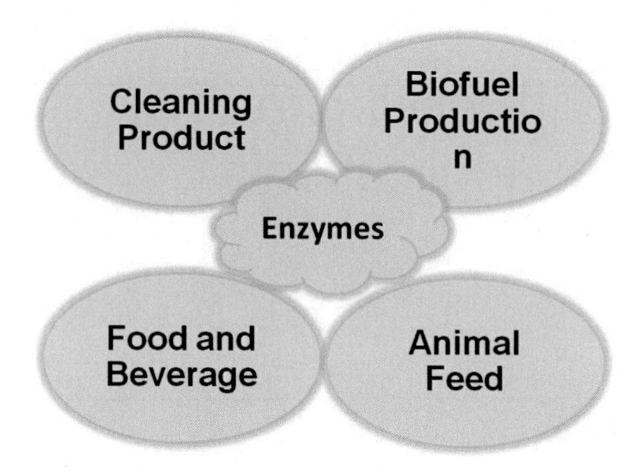

FIGURE 9.1

Important industrial sectors with enzyme applicability.

9.2 Enzymes in industries

There are innumerable advantages of enzyme usage, yet the sensitivity poses a problem in the expansion of its applicability to a variety of demanding industrial transformations. To combat the sensitivity of enzymes, it is immobilized using a conventional manner or informs using nanotechnology. The conventional manner of immobilization is adsorption, entrapment, cross-linking, and covalent linkages. Fig. 9.2 explains the major advantages of the use of immobilized enzymes in the industrial sector.

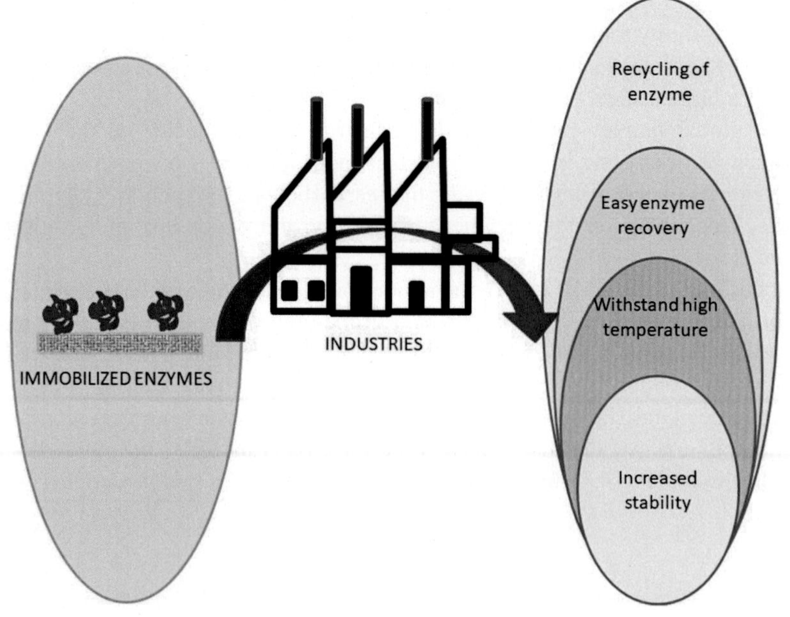

FIGURE 9.2

Advantages offered by immobilized enzymes used in industries.

Table 9.1 Comparative properties of free and immobilized enzyme forms.

Native enzyme	Immobilized enzyme
Activity is more	Reduced activity (Homaei et al., 2013)
Use is economic	Cost of immobilization carrier increases
Disposition after usage is not encountered	Incineration after complete exhaustion of activity of immobilized enzyme (Basso and Serban, 2019)
Higher reaction rate is offered	Reaction rate slows down (Sara et al., 2020)
No issues like fouling	Fouling issues may occur (Luo et al., 2020)

Industrial processes do not offer much choice as their requirement is very stringent. The economic perspective of the process is the main selection force whether free or immobilized enzyme can be used. Heterogeneous catalysis also offers a widespread avenue to be selected by industry sectors. Heterogeneous catalysis helps in the reuse of enzymes, a long period of enzyme activity can be enjoyed, and it offers expanded applicability to a number of unrelated processes.

Immobilization technique also accompanies certain disadvantages if selected for industrial processes. Table 9.1 enlists the comparative features of free immobilized enzymes.

The immobilization depends on the activity of the enzyme retained after its attachment to the selected source of immobilization or the carrier. The km and the Vmax are kinetic parameters analyzed to judge the biotransformation efficacy. The km value indicates the enzyme-substrate affinity, and Vmax indicates the biotransformation efficiency. These parameters have a great significance while selection of the enzyme by the industrial sector (Carrillo et al., 2010).

9.3 Common strategies for enzyme immobilization

9.3.1 Use of carriers

A solid is selected upon which the enzyme is bound. As it is a surface phenomenon the adsorption is physical or chemical (Zhang et al., 2013). This offers the benefit of enzyme binding on the support using nonspecific reactions. Specific interactions are used to support the binding of the solid and enzyme in case covalent binding is involved. Table 9.2 provides an insight into the range of support materials used for the immobilization of industrially important enzymes.

9.3.2 Covalent linkages

The concept of immobilization of enzymes using covalent linkage is quite historic. Since 1977, scientists are trying to expand enzyme applications to varied industries. The immobilization studies revealed the internal stoichiometry of the enzyme and the proximation chemistry with the substrate. The solid support used for enzyme immobilization was referred as an adduct. As the support is mechanically stable, the recovery of the enzyme after a rigorous industrial process was visioned to be promising (Falb, 1977).

Table 9.2 Various enzymes immobilized using diverse support materials.

Support used	Enzyme immobilized	References
Electrospun polyacrylonitrile glycopolymer nanofiber	Catalase	Li et al. (2012)
Carbon nanotubes	Lysozyme Amylase	Neupane et al. (2019)
Hydrophilic polymer with nanogels	Pancreatic lipase	Ji et al. (2016)
Chitosan	Chymotrypsin	Adriano et al. (2008)
Functionalized electrospun nanofibers from poly (AN-co-MMA)	β-Galactosidase	El-Aassar (2013)

This alternative method inculcates chemical reactions that modify the solid surface or the adduct and the enzyme so that there can be an efficient binding. The covalent binding ensures an efficient immobilization with negligible loss of enzyme and fall-out rate. However, the chemical reactions that are used for linkage offer a detrimental effect on enzyme as the chemicals are executing their reactions at extremes of environmental condition. Many instances are observed where the chemicals have undesirably reacted with amino acids participating in the active site resulting in decreased enzyme activity of the immobilized form (Brena et al., 2013).

Immobilization using nanoparticles is the recent trend to increase the efficiency of immobilization (Nandini et al., 2020; Joshi et al., 2019, 2021; Geetha et al., 2021; Yu et al., 2021). The increased efficiency is thought to be an attribute of a high surface-to-volume ratio (Vaghari et al., 2016; Singh et al., 2019; Gherardi et al., 2019; An et al., 2020).

9.3.3 Entrapment

This is yet another popular approach for immobilization. In this method, the enzyme is mixed with porous material so that it remains in the matrix boundary yet it really catalyzes substrate to product through the pores of entrapping material. The entrapment material can be calcium alginate (Fraser and Bickerstaff, 1997), sol-gels (Drozdov et al., 2016), polymers (Lyu et al., 2021), nanomaterials (Jogaiah et al., 2019; Arabacı et al., 2021), and so on.

9.4 Some important industrial enzymes

9.4.1 Fructosyltransferase

Fructosyltransferase enzymes use sucrose as a substrate and synthesize oligosaccharides of fructose by transferring fructose molecules to the growing chain. The

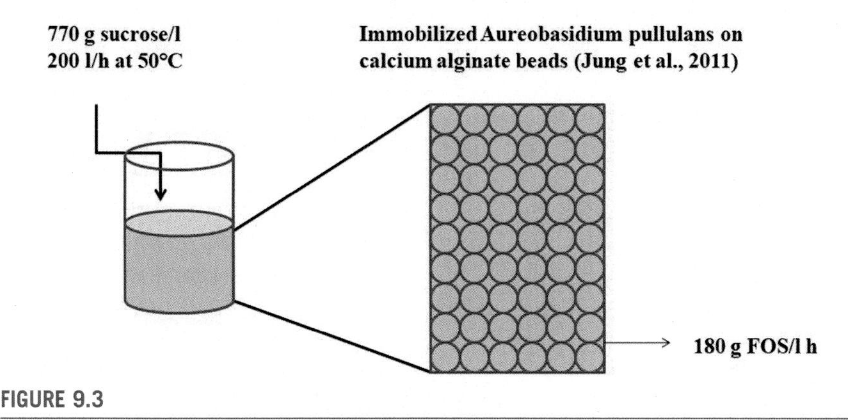

FIGURE 9.3

Large-scale production of fructooligosaccharides.

industrial production of fructooligosaccharides (FOS) has been successfully implemented by Meiji Seika industry using immobilized *Aspergillus niger* cells in calcium alginate. The second company Cheil Food and Chemicals (Seoul Korea) that successfully produced FOS on an industrial scale using immobilized *Aspergillus pullulans* in a calcium alginate gel (Sánchez-Martínez et al., 2020).

Jung et al. (2011) successfully scaled up the process of FOS production using fructosyltransferase produced by Continuous production of FOS by *Aureobasidium pullulans* immobilized on calcium alginate beads. The packed bed bioreactor was experimented at the initial stage with sucrose as substrate, 50°C temperature using cells in calcium alginate beads that yielded 180 g FOS per hour per liter of medium. The activity of the beads was stable for 100 days. The scaling up of the process was successful for the higher-capacity reactor. The process is diagrammatically represented in Fig. 9.3.

Fructosyltransferase has been experimented in all magnitude for optimum production of FOS at the industrial level (Kim et al., 1996). The process is either dependent on free enzyme biotransformation or the use of an immobilized enzyme or free cells as a source of enzyme. The free enzyme biotransformation is a batch process with a fixed incubation period and a specific volume of the substrate being transformed to a fixed amount of product. The use of free enzyme has its own disadvantages hence continuous production of FOS is also experimented using immobilized enzymes or cells (Burghardt et al., 2019). The use of immobilized enzymes or cells confers the process a preferred edge of recyclability of the enzyme, high activity inside the reactor vessel, and high degree of product yield as compared to free enzyme.

The immobilized enzyme application in industries has excelled the use of free enzymes also due to ease in downstream processing, the improved stability of the enzyme to function in the industrial process environment, the feasibility of continuous operations, and above all reusability of the immobilized form of the enzyme. The immobilization provides the enzyme a shield against the influence of harsh

Table 9.3 Materials used for immobilization of fructosyltransferase.

Materials	References
Methacrylamide-based polymeric beads	Chung et al. (1997)
Porous silica	Hayashi et al. (1991, 1992, 1993)
Acrylic carrier	Onderková et al. (2007)
Gellan gum	Daniel et al. (2018)
Polymethacrylate	Ghazi et al. (2005)
Macroporous beads	Tanriseven and Aslan (2005)
Calcium alginate	Oscar et al. (2019)
Amberlite	Csanadi and Sisak (2008)
Chitosan-coated magnetic nanoparticles	Chen et al. (2014), Yu et al. (2021)
Biodegradable polymers by Electrospinning Sepabeads	Gabrielczyk et al. (2018)
Calcium alginate	Lim et al. (2007a,b)
Covalent binding to epoxy functional groups	Burghardt et al. (2019)
Chitosan	Nandeesh Kumar et al. (2008), Bersaneti et al. (2019), Jogaiah et al. (2020)

operational parameters like temperature, pressure and autolytic effect of organic solvents (Fernandes, 2010; Mateo et al., 2007a,b; Sheldon, 2007).

For the enzyme to endure in the previously mentioned process conditions and down streaming processes, suitable immobilization techniques must be used. The general methods used are cross-linking of the enzyme to derive a carrier-free macromolecule, microencapsulation, and the use of a solid carrier (Khanvilkar et al., 2016). Table 9.3 provides information regarding various materials used for the immobilization of fructosyltransferase for the production of oligosaccharides.

9.4.2 Lipases

Lipases offer striking functional and structural attributes that expand their applicability in the industrial sector. To inculcate and improvise the stability of the enzyme, immobilization has been extensively practiced. Lipases are enzymes classified under Class 3 Hydrolases. These are acknowledged for their catalytic performance in hydrolysis, esterification, alcohol lysis, acidolysis, and aminolysis (Gandhi et al., 2000; Stergiou et al., 2013). Fig. 9.4 explains the major attributes of lipase rendering it as one of the most popular industrial enzymes.

Lipases exhibit actions on a broad range of substrates desiring different types of catalytic modifications. The diverse industries that are now dependent on lipases based biotransformation are as follows:

- Dairy and food industries
- Flavor and aroma components
- Leather and detergent industry
- Medical application
- Agrochemical and pharmaceutical industries

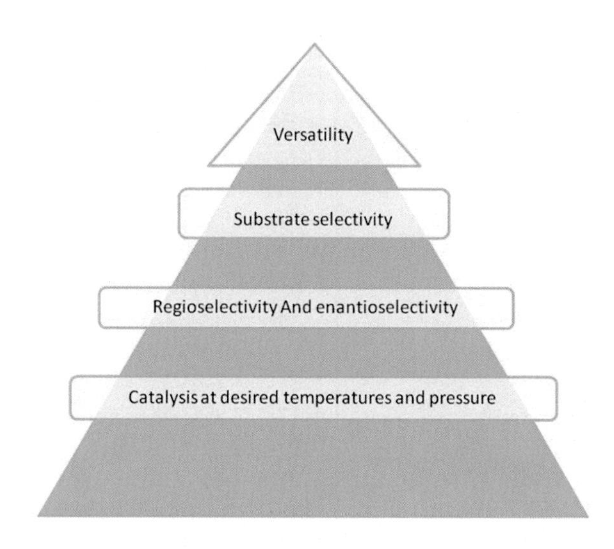

FIGURE 9.4

Versatile nature of lipase.

There are varieties of immobilization techniques for lipase. The choice depends on factors like the nature of enzymes, reaction conditions, solvents used, and the type of products formed. The selection in turn decides the type of bioreactor used. The various types of bioreactors employed in the use of immobilized lipase are shown in Fig. 9.5.

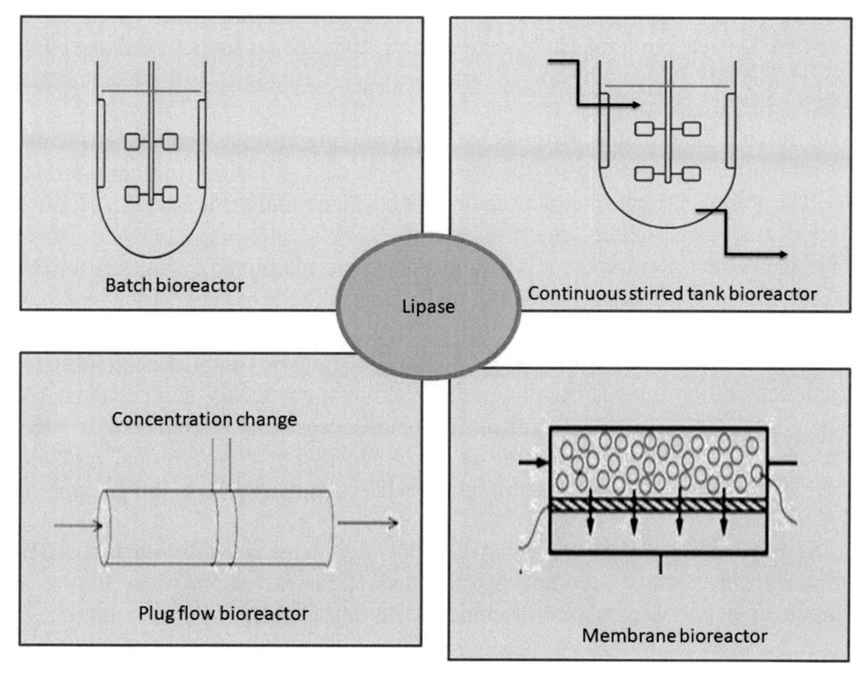

FIGURE 9.5

Various types of bioreactors using immobilized lipase.

Table 9.4 Carriers successfully used for lipase immobilization.

Carrier used	References
Alumina	Padmini et al. (1993)
Silica	Deon et al. (2020)
Celite	Kumar and Kanwar (2011)
Ceramics	Mulinari et al. (2020)
Metal oxides	Ali et al. (2018)
Graphite oxide	Farmakes et al. (2020)
Nonporous glass	Ghasemi et al. (2019)
Sepharose	Facchini et al. (2018)
Sephadex	Kaja et al. (2018)
Cellulose	Alnoch et al. (2020)
Zeolites	Calgaroto et al. (2011)
Modified polylactic acid	Li et al. (2020)
Polypropylene	Mokhtar et al. (2020)
Polystyrene	Li et al. (2010)
Nylons	Pahujani et al. (2008)

Among all the methods of immobilization, the most effective immobilization for lipases is the use of adsorption in combination with nonaqueous solvents. The use of nonaqueous solvents mitigates the enzyme desorption to a minimum level on grounds of poor solubility of the enzyme in the selected solvents. Thus, the adsorbed lipase finds great applications in industries using nonaqueous solvents (Kumar et al., 2016; Haque and Prabhu, 2016). The successful carriers used for these techniques are listed in Table 9.4.

Apart from the previous methods, covalent binding also has been experimented as it minimizes the problem of substrate and product diffusion to a considerable level. The common supports experimented for covalent binding are starch, cellulose, stainless steel, metal oxides, and many more.

Physical enzyme entrapment is also an alternative attempted for lipase immobilization. The trapping materials experimented fall in both organic as well as inorganic polymers. Both the materials offer their own advantages and drawbacks. The inorganic carriers by virtue of their material strength confer the immobilized enzyme the ability to resist high temperatures, harsh chemicals, and solvents. The efficient methanolysis of sunflower oil has been successfully attempted in a three-stage packed bioreactor by Tran et al. explained in Fig. 9.6. Physical pressure, growth of contaminants, and ease the process till product recovery and enzyme recovery as well.

The drawback of entrapment is the diffusion of enzyme and substrate that can be minimized but cannot be completely avoided. Table 9.5 summarizes important membrane reactor systems used in immobilized lipase-mediated transformation.

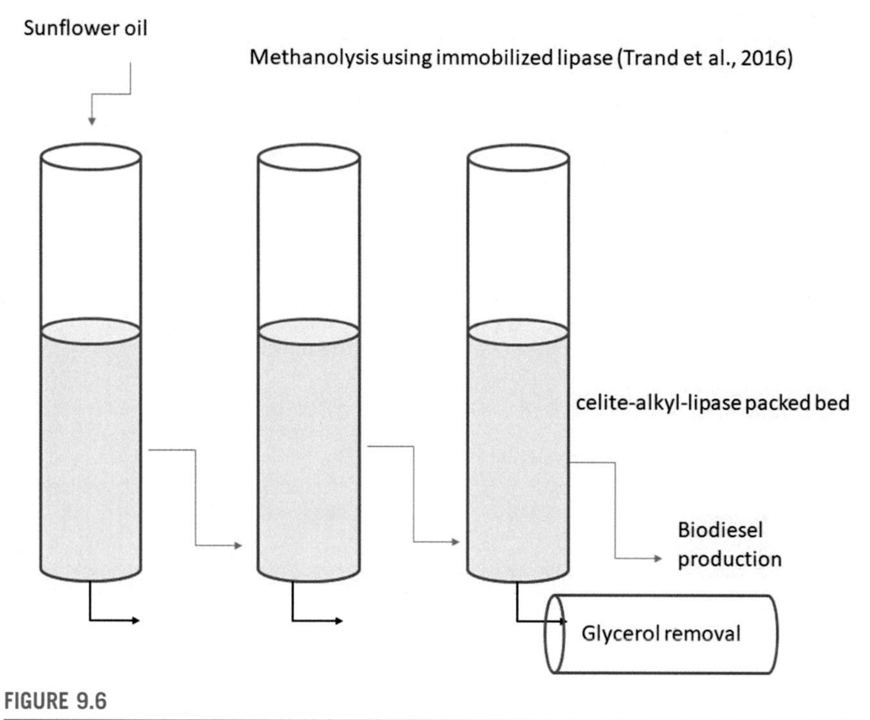

FIGURE 9.6

Biodiesel production using packed bed bioreactor of lipase.

9.4.3 Proteases

The protease enzyme also captures a respectable percent of the enzyme market. It is one of the most expensive biomolecules that have exuberant industrial demand. This has led the researchers to focus on methodologies that can enhance the reusability of protease in the conditions prevailing in the industrial process demanding the activity of the enzyme. All the factors which prove to be vital while immobilizing any other enzyme should be considered applicable for protease immobilization also (Liese and Hilterhaus 2013). The immobilization strategy of protease focuses on the improvisation of the immobilization process by increasing the number of attachment sites between the enzyme and the support. The multiple covalent bonding between the enzyme and the support is explained diagrammatically in Fig. 9.7.

The support selected for multiple covalent attachments should have specific features to qualify to be used. The salient features required for multicovalent linkages are given in Fig. 9.8.

Physical adsorption is also employed for immobilization of the protease. The process involves harsh conditions favoring adsorption and resulting in the inactivation of active enzyme functional groups (Garcia-Galan et al., 2011).

Although this mechanism is not so effective in enhancing the enzyme features yet has a positive influence on polymeric enzymes by imparting them strong binding (Gupta and Chaudhury, 2007).

Table 9.5 Immobilized lipase applications using bioreactors.

Bioreactors	Reaction catalyzed	Important features
Emulsion membrane reactors (Giorno et al., 2003)	Optical pure (S) naproxen from racemic naproxen methyl ester	In continuous systems, these membranes show features of low-pressure drop Short residence time High operational stability A low external and internal diffusional resistances
Microporous hydrophobic hollow fiber bioreactor (Hoq et al., 1985)	Olive oil hydrolysis	Lipase was immobilized at the water–glycerol side of the membrane
Hydrophobic reactor system Malcata et al. (1991) Mercon et al. (1997) Balcao and Malcata (1997)	Hydrolysis of glycerides of butter oil Babassu oil hydrolysis Acidolysis	Enzyme has hydrophilic center Essential for membrane activity. Emulsifiers influence the interaction between substrate and enzyme.
Hydrophilic hollow-fiber Membrane system (Okobira et al., 2015).	Esterification reaction Polymer using amino ethanol group was used. Positive charge repulsion during AE groups proximity promotes immobilization of lipase in layered form improving the catalytic efficiency	Has a reduced enzyme desorption as an outcome of insolubility in the organic phase. The catalytic activities per membrane area are independent membrane dimensions. The catalysis depends on the quantity of enzymes loaded that are high for this type of membrane.

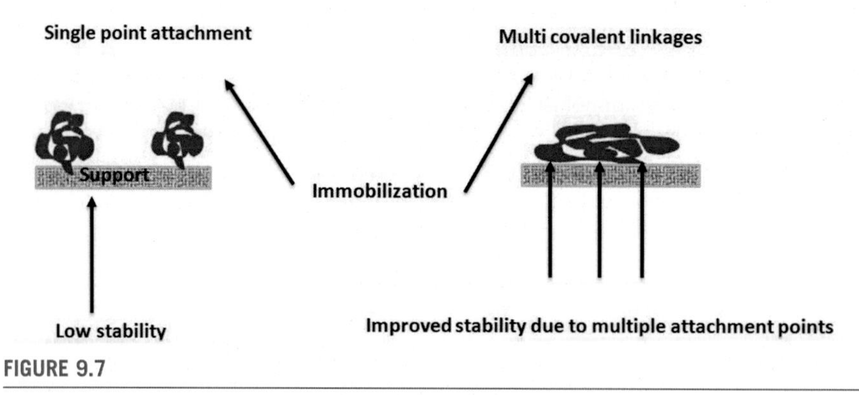

FIGURE 9.7

Multiple attachments between enzyme and the support.

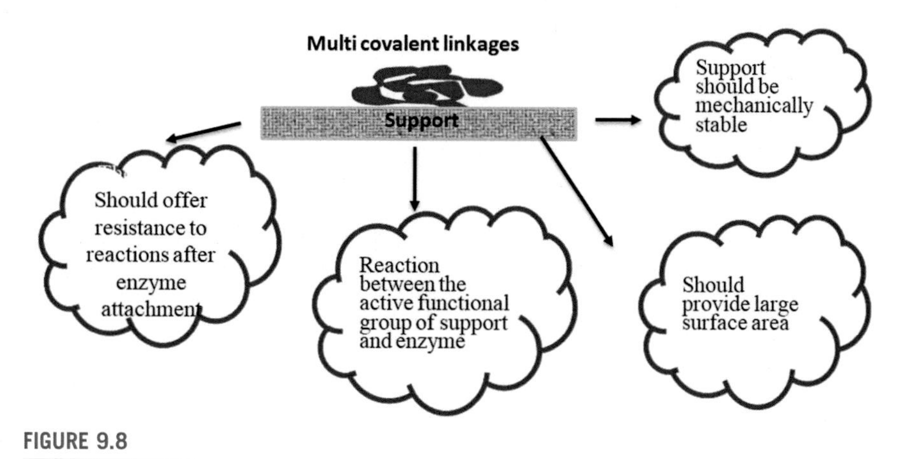

FIGURE 9.8

Salient features of support to be selected for multicovalent attachment strategy.

Proteases are experimented with innovative approaches for enhanced functionalities, and to delimit the restrictions of fall out of enzymes, polymer-based modifications are attempted for increased enzyme size. The false increase in size provides better attachment, which is also achieved by using polymers to create enzyme aggregates (Hiroshi et al., 2018).

9.5 Biocatalytic membranes

Recent advances in enzyme immobilization strategies include the use of biocatalytic membranes. The use of membranes in biotechnological enzyme application has successfully amalgamated the three important requisites of the industrial sector, that is, maintenance of enzyme activity, recyclability, and ease in the separation of the product. The process also offers a separate space for effective enzyme–substrate interaction, enzyme as well as product recovery. However, this process requires amendments in improving enzyme potential and cost reduction. The industrialization of biocatalytic membranes is facing challenges related to cost, enzyme stability, and enzyme performance that call an urgent scientific address (Luo et al., 2020).

The membrane-based bioreactors are successfully attempted for enzyme protease (Das, 2014). The reactor design offers a reaction space limited by a membrane. This is an attempt to mimic the selectivity potential of the natural cell membrane. The biomembrane selectively permits only product to transport across it whereas the substrate, enzyme, and the long peptides are retained in the mixture. This method offers the advantage of complete enzyme recovery. The advantages of this method are as follows:

Complete retainment of enzyme.
Full capacity recycling.

Effective reduction in cost.
Pore size of the membrane can be selected according to the process requirement.
Protease is retained active even at the end of the process in contrast to the conventional approach where deactivation of the enzyme is the only choice for reaction termination.

9.6 Conclusion

The conventional methods of enzyme immobilization are effective even in the present scenario as calcium alginate-immobilized fructosyltransferase has been adopted for industrial production of fructooligosaccharides by two pioneering companies. Although, immobilization is successfully attempted by methods such as adsorption, covalent linking, multipoint linkages, entrapment, and use of carrier, yet all of them offer major limitations to be successfully implemented for a satisfactory product yield at the industrial level.

Bioreactors should be tailor designed according to the immobilization process selected and the issues of mechanical stress, pressure, and solvents have to be optimized. Approaches based on nonaqueous medium, organic solvents, novel materials, and nanoparticles are some of the promising alternatives to make immobilization a wonderful industrial tool with enzyme savvy features rendering the process economical and the product affordable.

References

Adriano, W.S., Mendoca, D.B., Rodrigues, D.S., 2008. Improving the properties of chitosan as support for the covalent multipoint immobilization of chymotrypsin. Biomacromolecules 9, 2170–2179.

Ali, Z., Li, T., Khan, M., Ali, N., Zhang, Q., 2018. Immobilization of lipase on iron oxide organic/inorganic hybrid particles: a review. Rev. Adv. Mater. Sci. 53, 106–117.

Alnoch, R.C., dos Santos, L.A., de Almeida, J.M., Krieger, N., Mateo, C., 2020. Recent trends in biomaterials for immobilization of lipases for application in non-conventional media. Catalysts 10, 697. https://doi.org/10.3390/catal1006069.

An, J., Li, G., Zhang, Y., Zhang, T., Liu, X., Fei, G., Peng, M, He, Y., Fan, H., 2020. Recent advances in enzyme-nanostructure biocatalysts with enhanced activity. Catalyst 10, 338. https://doi.org/10.3390/catal10030338.

Arabacı, N., Karaytug, N., Demirbas, N., Ocsoy, I., Katı, A., 2021. In: Srivastava, N., Srivastava, M., Mishra, P.K., Gupta, V.K. (Eds.), Nanomaterials for enzyme immobilization green synthesis of nanomaterials for bioenergy applications, first ed. John Wiley & Sons Ltd.

Balcao, V.M., Malcata, F.X., 1998. On the performance of a hollow-fiber bioreactor for acidolysis catalyzed by immobilized lipase. Biotechnol. Bioeng. 60, 114–123.

Basso, A., Serban, S., 2019. Industrial applications of immobilized enzymes—a review. Mol. Catal. 479.

Bersaneti, G.T., Baldo, C., Antonia, M., Celligoi, A.P., 2019. Immobilization of levansucrase: strategies and biotechnological applications. J. Chil. Chem. 64. https://doi.org/10.4067/s0717-97072019000104377.

Brena, B., González-Pombo, P., Batista-Viera, F., 2013. Immobilization of enzymes: a literature survey. In: Guisan, J.M. (Ed.), Immobilization of Enzymes and Cells, Methods in Molecular Biology, third ed., vol. 1051. © Springer Science+Business Media, New York. https://doi.org/10.1007/978-1-62703-550-7_2.

Burghardt, J.P., Baas, M., Gerlach, D., Czermak, P., 2019. Two-step production of neofructo-oligosaccharides using immobilized heterologous *Aspergillus terreus* 1F-fructosyltransferase expressed in *Kluyveromyces lactis* and native *Xanthophyllomyces dendrorhous* G^6-fructosyltransferase. Catalysts 9 (8), 673. https://doi.org/10.3390/catal9080673.

Calgaroto, C., Scherer, R.P., Calgaroto, S.J., Oliveira, V., de Oliveira, D., Pergher, S.B.C., 2011. Immobilization of porcine pancreatic lipase in zeolite MCM 22 with different Si/Al ratios. Appl. Catal. A Gen. 394, 101–104.

Carrillo, N., Ceccarelli, E.A., Roveri, O.A., 2010. Usefulness of kinetic enzyme parameters in biotechnological practice. Biotechnol. Genet. Eng. Rev. 27 (1), 367–382. https://doi.org/10.1080/02648725.2010.10648157.

Chen, S.C., Sheu, D.C., Duan, K.J., 2014. Production of fructooligosaccharides using β-fructofuranosidase immobilized onto chitosan-coated magnetic nanoparticles. J. Taiwan Inst. Chem. Eng. 45 (4), 1105–1110.

Chung, J.C., Lee, W.C., Kow, D.C., Duan, J., 1997. Immobilization of β-fructofuranosidases from *Aspergillus* on methacrylamide based polymeric beads for production of fructooligosaccharides. Biotechnol. Prog. 13, 577–582.

Csanadi, Z., Sisak, C., 2008. Production of short chain fructooligosaccharides. Hung 1, 23–26.

Daniel, H., Valdeón, Paula, Z., Araujo, Mirta, D.N., Perotti, I., 2018. Immobilization of fructofuranosidase from *Aureobasidium* sp. onto TiO_2 and its encapsulation on gellan gum for FOS production. Int. J. Chem. React. Eng. https://doi.org/10.1515/ijcre-0135.

Das, R., 2014. Proteases in membrane bioreactor. In: Drioli, E., Giorno, L. (Eds.), Encyclopedia of Membranes. Springer, Berlin, Heidelberg. https://doi.org/10.1007/978-3-642-40872-4_1855-1.

Deon, M., Ricardi, N.C., de Andrade, N.C., Hertz, P.F., Nicolodi, S., Costa, T.M.H., Bussamara, E., Valmir Benvenutti, E.V., de Menezes, E.W., 2020. Designing a support for lipase immobilization based on magnetic, hydrophobic, and mesoporous silica. Langmuir 36 (34), 10147–10155. https://doi.org/10.1021/acs.langmuir.0c01594.

Drozdov, A.S., Shapovalova, O.E., Ivanovski, V., Avnir, D., Vinogradov, V.V., 2016. Entrapment of enzymes within sol–gel-derived magnetite. Chem. Mater. 28 (7), 2248–2253. https://doi.org/10.1021/acs.chemmater.6b00193.

El-Aassar, M.R., 2013. Functionalized electrospun nanofibers from poly (AN-co-MMA) for enzyme immobilization. J. Mol. Catal. B Enzym. 85–86, 140–148.

Facchini, F.D.A., Pereira, M.G., Vici, A.C., Filice, M., Pessela, B.C., Guisan, J.M., Fernandez-Lorente, G., de Moraes Polizeli, M.D.L.T., 2018. Immobilization effects on the catalytic properties of two fusarium verticillioides lipases: stability, hydrolysis, transesterification and enantioselectivity improvement. Catalysts 8, 84. https://doi.org/10.3390/catal8020084.

Falb, R.D., 1977. Covalent linkage: I. Enzymes immobilized by covalent linkage on insolubilized supports. In: Chang, T.M.S. (Ed.), Biomedical Applications of Immobilized

Enzymes and Proteins. Springer, Boston, MA. https://doi.org/10.1007/978-1-4684-2610-6_2.

Farmakes, J., Schuster, I., Overby, A., Alhalhooly, L., Lenertz, M., Li, Q., Ugrinov, A., Choi, Y., Pan, Y., Fernandes, P., 2010. Enzymes in food processing: a condensed overview on strategies for better biocatalysts. Enzym. Res. 1–19.

Fernandes, P., 2010. Enzymes in food processing: a condensed overview on strategies for better biocatalysts. Enzyme Res. https://doi.org/10.4061/2010/862537. Article ID 862537.

Fraser, J.E., Bickerstaff, G.F., 1997. Entrapment in calcium alginate. In: Bickerstaff, G.F. (Ed.), Immobilization of Enzymes and Cells, Methods in Biotechnology, vol. 1. Humana Press, p. 61. https://doi.org/10.1385/0-89603-386-4.

Gabrielczyk, J., Duensing, T., Buchholz, S., Schwinges, A., Jordening, H.J., 2018. A comparative study on immobilization of fructosyltransferase in biodegradable polymers by electrospinning. Appl. Biochem. Biotechnol. 185, 847–862. https://doi.org/10.1007/s12010-018-2694-6.

Gandhi, N.N., Patil, N.S., Sawant, S.B., Joshi, J.B., Wangikar, P.P., Mukesh, D., 2000. Lipase-Catalyzed Esterification. Catalysis Reviews 42 (4), 439–480.

Garcia-Galan, C., Berenguer-Murcia, A., Fernandez-Lafuente, R., Rodrigues, R.C., 2011. A review potential of different enzyme immobilization strategies to improve enzyme performance. Adv. Synth. Catal. 353, 2885–2904.

Geetha, N., Bhavya, G., Abhijith, P., Shekhar, R., Dayananda, K., Jogaiah, S., 2021. Insights into nanomycoremediation: secretomics and mycogenic biopolymer nanocomposites for heavy metal detoxification. J. Hazard Mater. 409, 124541. https://doi.org/10.1016/j.jhazmat.2020.124541.

Ghasemi, S.S., Heidary, M., Bozorgi-Koushalshahi, M., Habibi, Z., Gherardi, F., Turyanska, L., Ferrari, E., Weston, N., Fay, M.W., Colston, B.J., 2019. ACS Appl. Bio. Mater. 2 (11), 5136–5143. https://doi.org/10.1021/acsabm.9b00802.

Ghazi, I., Segura, A., Fernández-Arrojo, L., Alcalde, M., Yates, M., Rojas-Cervantes, M., Plou, F., Ballesteros, A., 2005. Immobilisation of fructosyltransferase from Aspergillus aculeatus on epoxy-activated Sepabeads EC for the synthesis of fructo-oligosaccharides. J. Mol. Catal. B Enzym. 35, 19–27.

Gherardi, F., Turyanska, L., Ferrari, E., Weston, N., Fay, M.W., Colston, B.J., 2019. Immobilized enzymes on gold nanoparticles: from enhanced stability to cleaning of textile heritage. ACS. Appl. Bio Mater. 2 (11), 5136–5143.

Giorno, L., Li, N., Drioli, E., 2003. Use of stable emulsion to improve stability, activity, and enantioselectivity of lipase immobilized in a membrane reactor. Biotechnol. Bioeng. 84, 677–685.

Gupta, R., Chaudhury, N.K., 2007. Entrapment of biomolecules in sol–gel matrix for applications in biosensors: Problems and future prospects. Biosens. Bioelectron. 22, 2387–2399.

Haque, N., Prabhu, N.P., 2016. Lid dynamics of porcine pancreatic lipase in non-aqueous solvents. Biochim. Biophys. Acta Gen. Subj. 1860, 2326–2334.

Hayashi, S., Hayashi, T., Kinoshita, J., et al., 1992. Immobilization of b-fructofuranosidase from *Aureobasidium* sp. ATCC 20524 on porous silica. J. Ind. Microbiol. Biotechnol. 9, 247–250.

Hayashi, S., Ito, K., Nonoguchi, M., et al., 1991. Immobilization of a fructosyl-transferring enzyme from *Aureobasidium* sp. on shirasuporous glass. J. Ferment. Bioeng. 72, 68–70.

Hayashi, S., Matsuzaki, K., Inomata, Y., et al., 1993. Properties of *Aspergillus japonicus* b-fructofuranosidase immobilized on porous silica. World J. Microbiol. Biotechnol. 9, 216–220.

Hiroshi, Y., Yuhei, K., Masaya, M., 2018. Techniques for preparation of cross-linked enzyme aggregates and their applications in bioconversions. Catalysts 8, 174. https://doi.org/10.3390/catal8050174.

Homaei, A.A., Sariri, R., Vianello, F., Stevanato, R., 2013. Enzyme immobilization: an update. J. Chem. Biol. 6 (4), 185–205. https://doi.org/10.1007/s12154-013-0102-9.

Hoq, M.M., Koike, M., Yamane, T., Shimizu, S., 1985. Continuous hydrolysis of olive oil by lipase in microporous hydrophobic hollow fiber bioreactor. Agric. Biol. Chem. 49, 3171–3178.

Ji, X., Liu, J., Liu, L., Zhao, H., 2016. Enzyme-polymer hybrid nanogels fabricated by thiol-disulfide exchange reaction. Colloids Surf. B Biointerfaces 148, 41–48.

Jogaiah, S., Kurjogi, M., Abdelrahman, M., Nagabhushana, H., Tran, L.-S.P., 2019. *Ganoderma applanatum*-mediated green synthesis of silver nanoparticles: structural characterization and in vitro and in vivo biomedical and agrochemical properties. Arab. J. Chem. 12, 1108–1120. https://doi.org/10.1016/j.arabjc.2017.12.002.

Jogaiah, S., Praveen, S., De Britto, S., Konappa, N., Udayashankar, A.C., 2020. Exogenous priming of chitosan induces upregulation of phytohormones and resistance against cucumber powdery mildew disease is correlated with localized biosynthesis of defense enzymes. Int. J. Biol. Macromol. 162, 1825–1838. https://doi.org/10.1016/j.ijbiomac.2020.08.124.

Joshi, S.M., De Britto, S., Jogaiah, S., 2021. Myco-engineered selenium nanoparticles elicit resistance against tomato late blight disease by regulating differential expression of cellular, biochemical and defense responsive genes. J. Biotechnol. 325, 196–206. https://doi.org/10.1016/j.jbiotec.2020.10.023.

Joshi, S.M., De Britto, S., Jogaiah, S., Ito, S., 2019. Mycogenic selenium nanoparticles as potential new generation broad spectrum antifungal molecules. Biomolecules 9 (9), 419. https://doi.org/10.3390/biom9090419.

Jung, K.H., Bang, S.H., Oh, T.K., Park, H.J., 2011. Industrial production of fructooligosaccharides by immobilized cells of Aureobasidium pullulans in a packed bed reactor. Biotechnol. Lett. 33 (8), 1621–1624. https://doi.org/10.1007/s10529-011-0606-8.

Kaja, B., Lumor, S., Besong, S., Taylor, B., Ozbay, G., 2018. Investigating enzyme activity of immobilized *Candida rugosa* Lipase. J. Food Qual. 9. https://doi.org/10.1155/2018/1618085. Article ID 1618085.

Khanvilkar, A.M., Ranveer, R.C., Sahoo, A.K., 2016. Carrier materials for encapsulation of bio-active components of food. Int. J. Pharm. Sci. Rev. Res. Article No. 14. 40, 62–73.

Kim, M.-H., In, M.-J., Cha, H.J., Yoo, Y.J., 1996. An empirical rate equation for the fructooligosaccharide-producing reaction catalyzed by β-fructofuranosidase. J. Ferment. Bioeng. 82, 458–463.

Kumar, A., Kanwar, S.S., 2011. Synthesis of ethyl ferulate in organic medium using celite-immobilized lipase. Bioresour. Technol. 102, 2162–2167.

Kumar, A., Dhar, K., Kanwar, S.S., et al., 2016. Lipase catalysis in organic solvents: advantages and applications. Biol. Proced. Online 18, 2. https://doi.org/10.1186/s12575-016-0033-2.

Li, Y., Gao, Y., Wei, F., Qu, Z., Jian-Bo, M., Guang-Hui, Z., Qing, W., 2010. Pore size of macroporous polystyrene microspheres affects lipase immobilization. J. Mol. Catal. B Enzym. 66, 182–189. https://doi.org/10.1016/j.molcatb.2010.05.007.

Li, Y., Jing, Q., Whjite, B., Christopher, Williams, G., Wu, Xian, J., Zhu, Li, M., 2012. Electrospun polyacrylonitrile-glycopolymer nanofibrous membranes for enzyme immobilization. J. Mol. Catal. B Enzym. 76, 15–22.

Li, S., Zhao, S., Hou, Y., Chen, G., Chen, Y., Zhang, Z., 2020. Polylactic Acid (PLA) Modified by Polyethylene Glycol (PEG) for the immobilization of lipase. Appl. Biochem. Biotechnol. 190, 982–996.

Liese, A., Hilterhaus, L., 2013. Evaluation of immobilized enzymes for industrial applications. Chem. Soc. Rev. 42, 6236–6249.

Lim, J., Lee, J., Kang, S., et al., 2007a. Studies on production and physical properties of neo-FOS produced by co-immobilized *Penicilliumcitrinum* and neo-fructosyltransferase. Eur. Food Res. Technol. 225, 457–462.

Lim, J.S., Lee, J.H., Kang, S.W., Park, S.W., Kim, S.W., 2007b. Studies on production and physical properties of neo-FOS produced by co-immobilized *Penicillium citrinum* and neo-fructosyltransferase. Eur. Food Res. Technol. 225, 457–462. https://doi.org/10.1007/s00217-006-0440-8.

Luo, J., Song, S., Zhang, H., Zhang, H., Zhang, J., Wan, Y., 2020. Biocatalytic membrane: go far beyond enzyme immobilization. Eng. Life Sci. 20 (11), 441–450. https://doi.org/10.1002/elsc.202000018.

Lyu, X., Gonzalez, R., Horton, A., Li, T., 2021. Review immobilization of enzymes by polymeric materials catalysts, 11, p. 1211. https://doi.org/10.3390/catal11101211.

Malcata, F.X., Hill, C.G., Amundson, C.H., 1991. Use of a lipase immobilized in a membrane reactor to hydrolyze the glycerides of butteroil. Biotechnol. Bioeng. 38853–38868.

Mateo, C., Grazu, V., Pessela, B., Montes, T., Palomo, J., Torres, R.T.R., Gallego, F.L., Fernandez-Lafuente, R., Guisan, J., 2007a. Advances in the design of new epoxy supports for enzyme immobilization-stabilization. Biochem. Soc. Trans. 35, 1593–1601.

Mateo, C., Palomo, J.M., Fernandez-Lorente, G., et al., 2007b. Improvement of enzyme activity, stability and selectivity via immobilization techniques. Enzym. Microb. Technol. 40, 1451–1463.

Mercon, F., Erbes, V.L., SantAnna, G.L., Nobrega, R., 1997. Lipase immobilized membrane reactor applied to babassu oil hydrolysis. Braz. J. Chem. Eng. 14, 1–11.

Mokhtar, N.F., Rahman, R.N.Z., Noor, N.D.M., Shariff, F.M., Ali, M.S.M., 2020. The immobilization of lipases on porous support by adsorption and hydrophobic interaction method. Catalysts 10, 744. https://doi.org/10.3390/catal10070744.

Mulinari, J., Oliveira, V., Hotza, D., 2020. Lipase immobilization on ceramic supports: an overview on techniques and materials. Biotechnol. Adv. 42, 107581.

Nandeesh Kumar, P.K., Jogaiah, S., Kini, K.R., Prakash, H.S., Niranjana, S.R., Shekar Shetty, H., 2008. Chitosan induced resistance to downy mildew in sunflower caused by *Plasmopara halstedii*. Physiol. Mol. Plant Pathol. 72, 188–194.

Nandini, B., Puttaswamy, H., Prakash, H.S., Adhikari, S., Jogaiah, S., Nagaraja, G., 2020. Elicitation of novel trichogenic-lipid nanoemulsion signaling resistance against pearl millet downy mildew disease. Biomolecules 10 (1), 25. https://doi.org/10.3390/biom10010025.

Neupane, S., Patnode, K., Li, H., Baryeh, K., Liu, G., Hu, J., Chen, B., Pan, Y., Yang, Z., 2019. Enhancing enzyme immobilization on carbon nanotubes via metal–organic frameworks for large-substrate biocatalysis. ACS Appl. Mater. Interfaces 11 (12), 12133–12141. https://doi.org/10.1021/acsami.9b01077.

Okobira, T., Matsuo, A., Matsumoto, A., Tanaka, H., Kai, T., Minari, K., Goto, C., Kawakita, M., Uezu, H., Kazuya, 2015. Enhancement of immobilized lipase activity by design of polymer brushes on a hollow fiber membrane. J. Biosci. Bioeng. 120. https://doi.org/10.1016/j.jbiosc.01.009.

Onderková, Z., Bryjak, J., Polakovič, M., 2007. Properties of fructosyltransferase from Aureobasidium pullulans immobilized on an acrylic carrier. Chem. Pap. 61, 5. https://doi.org/10.2478/s11696-007-0048-x.

Oscar, K.K., Bedzo, K.T., Lalitha, D., Gottumukkala, G.C., Johann, F.G., 2019. Amberlite IRA 900 versus calcium alginate in immobilization of a novel, engineered β-fructofuranosidase for short-chain fructooligosaccharide synthesis from sucrose. Biotechnol. Prog. 35, 3. https://doi.org/10.1002/btpr.2797.

Padmini, P., Rakshit, S.K., Baradarajan, A., 1993. Studies on immobilization of lipase on alumina for hydrolysis of ricebran oil. Bioprocess Eng. 9, 43–46. https://doi.org/10.1007/BF00389539.

Pahujani, S., Kanwar, S.S., Chauhan, G., Gupta, R., 2008. Glutaraldehyde activation of polymer Nylon-6 for lipase immobilization: enzyme characteristics and stability. Bioresour. Technol. 99, 2566–2570.

Sánchez-Martínez, M.J., Soto-Jover, S., Antolinos, V., et al., 2020. Manufacturing of short-chain fructooligosaccharides: from laboratory to industrial scale. Food Eng. Rev. 12, 149–172. https://doi.org/10.1007/s12393-020-09209-0.

Sara, A., Nathalia, S.R., Diego, C., Carmen, M.S., Yuliya, L., Luciana, G.R.B., Roberto, F.L., 2020. Effects of enzyme loading and immobilization conditions on the catalytic features of lipase from pseudomonas fluorescens immobilized on octyl-agarose beads. Front. Bioeng. Biotechnol. https://doi.org/10.3389/fbioe.2020.00036. https://www.frontiersin.org/article/10.3389/fbioe.2020.00036.

Sheldon, R.A., 2007. Enzyme immobilization: the quest for optimum performance. Adv. Synth. Catal. 349, 1289–1307.

Singh, K., Mishra, A., Sharma, D., Singh, K., 2019. Nanotechnology in enzyme immobilization: an overview on enzyme immobilization with nanoparticle matrix. Curr. Nanosci. 15, 234.

Stergiou, P.Y., Foukis, A., Filippou, M., Koukouritaki, M., Parapouli, M., Theodorou, L.G., Hatziloukas, E., Afendra, A., Pandey, A., Papamichael, E.M., 2013. Advances in lipase-catalyzed esterification reactions. Biotechnol. Adv. 31, 1846–1859.

Tanriseven, A., Aslan, Y., 2005. Immobilization of Pectinex Ultra SP-L to produce fructooligosaccharides. Enzyme Microb. Technol. 36, 550–554.

Vaghari, H., Jafarizadeh-Malmiri, H., Mohammadlou, M., Berenjian, A., Anarjan, N., Jafari, N., Nasiri, S., 2016. Application of magnetic nanoparticles in smart enzyme immobilization. Biotechnol. Lett. 38, 223–233. https://doi.org/10.1007/s10529-015-1977-z.

Yu, J., Wang, D., Geetha, N., Khawar, M., Mujtaba, M., Jogaiah, S., 2021. Current trends and challenges in the synthesis and applications of chitosan-based nanocomposites for plants. Carbohydr. Polym. 261, 117904. https://doi.org/10.1016/j.carbpol.2021.117904.

Zhang, D.H., Yuwen, L.X., Peng, L.J., 2013. Parameters affecting the performance of immobilized enzyme. J. Chem. https://doi.org/10.1155/2013/926248. ID 946248.

CHAPTER

Agroindustrial wastes for enzyme production

10

10.1 Introduction

Industrial enzymes have a significant history of their usage and acceptance in the industrial sector. The most preliminary enzyme amylase found basic applicability in industrial processes involving hydrolysis of starch. The sources of amylases were thoroughly experimented to render the enzyme production process economics. *Bacillus* sp. was the most desirable source for the economic production of amylase. The most commonly exploited species of Bacillus are *B. licheniformis*, *B. stearothermophilus,* and *B. amyloliquefaciens*. The industries that progressively became dependent on amylase are food, fodder, detergents, paper industry, etc. The Bacillus-derived amylase enzyme enjoys its supremacy over alternative amylase sources due to its heat resistant nature that makes it an industrially desirable molecule (Konsoula and Liakopoulou-Kyriakides, 2007; Babu et al., 2015).

10.2 Agroindustrial wastes: excellent raw material for production

The agriculture sector has gained limitless momentum since the beginning of the 21st century. Countries based on agriculture are creating greater and versatile trade in the international market. The variety of agricultural products available in the market is increasing due to the innovative transformation of traditional food to attractive forms with long shelf life. The demand for basic agricultural products such as cereals, sugar, and pulses has a constant demand in the market (Sudisha et al., 2005; Fang et al., 2018; Haider et al., 2019; Patil et al., 2020). With this increasing growth statistics of the agriculture sector, the proportional rise in the wastes generated by these practices is one of the sensitive issues of global concern. These wastes have immense exploitation potential and hence referred as agroindustrial wastes.

The agricultural practice provides a targeted crop or edible plant part that is marketed for the direct economic goal of agricultural practice. The agricultural products are transported to industries for processing and packaging intended for market sale. Both these platforms, that is, the agricultural fields and the linked industries generate

Protocols and Applications in Enzymology. https://doi.org/10.1016/B978-0-323-91268-6.00004-1

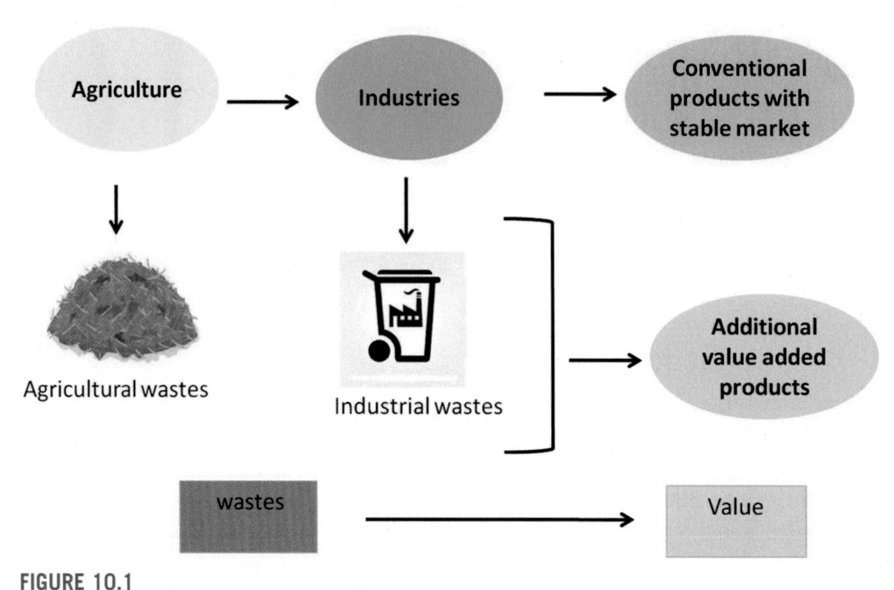

FIGURE 10.1

Generation of agroindustrial wastes.

a wide variety of wastes usually referred as agroindustrial wastes as shown in Fig. 10.1.

The regular agricultural activities that lead to the cultivation of crops leave field residues as parts of crops that are nonedible and left in the field. The agroindustrial wastes are further classified on basis of their sources as agricultural wastes and industrial wastes (Sangeetha et al., 2017). The wastes classified under each category are enlisted in Fig. 10.2.

10.3 Impact of agroindustrial wastes on environment: a global concern

As earlier stated, the growing agriculture sector and the linked industries produce waste materials or residues in huge and unimaginable quantities in the global scenario. Reports from the past decade indicate 998 million tonnes of agricultural waste production every year (Agamuthu, 2009; Bhavya et al., 2021). Interim report 2 by Govt of Mauritious has categorized the organic wastes to about 80% of the total solid wastes generated (Obi et al., 2016). Fig. 10.3 categorizes the quantity of wastes generated of each type.

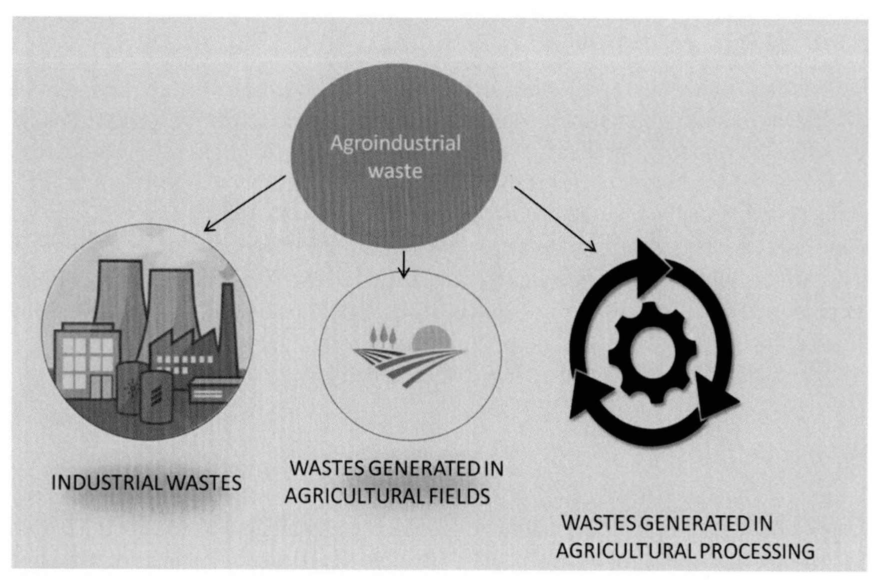

FIGURE 10.2

Classification of agroindustrial wastes.

Agroindustrial wastes generated during rice cultivation, wheat cultivation, and sugarcane processing are rice bran, wheat bran, wheat straw, and sugarcane bagasse that prove to be an excellent source of carbon for a variety of organisms. These carbon sources can serve as an excellent growth material for microbes involved in the production of industrially important enzymes (Geetha et al., 2021). The added advantage of such agroindustrial waste is its economy, that is, reduction in the overall process cost that makes its use industrially favorable. The utilization of agroindustrial waste for enzyme production has a double-sided benefit involving low-cost enzyme production and productive utilization of wastes rather than its disposal, which is an environmental concern.

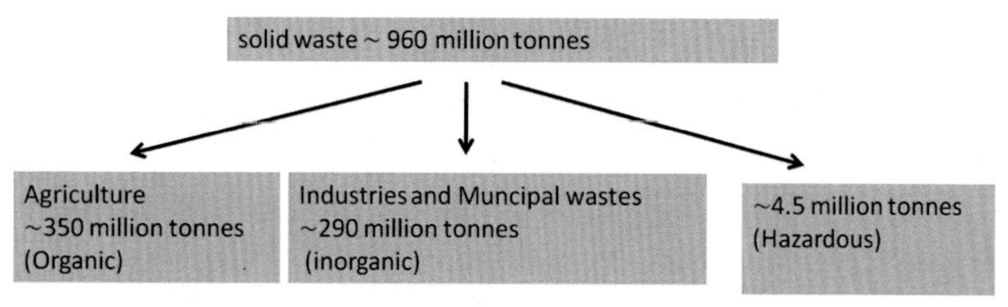

FIGURE 10.3

Production of wastes: a comparative overview (Pappu et al., 2007).

10.4 Waste to value

The enormous amounts of agroindustrial wastes generated prove to be a potential raw material for fermentations catered by industries for effective production of economically important products. This approach toward utilization of agroindustrial waste can be referred as a quadruple method involving effective utilization, economic, commercially favorable, and environment friendly (Sadh et al., 2018; Yu et al., 2021). The majority of wastes generated during agricultural practices are rich in lignocellulosic content (Dakshayani et al., 2015). This composition confers the plant material to be used as a source of renewable energy. The composition of lingo cellulosic material is cellulose, hemicellulose, and lignin with the occurrence at 35%–50%, 20%–35%, and 15%–25%, respectively.

10.4.1 Cellulose

The major component is the cellulose that is almost half of the lignocellulosic material. It is a carbohydrate with a linear structure of glucose monomers linked with the β-1,4-glycosidic bond. Cellulose is quite resistant to biological degradation in a natural environment. An enzyme cocktail is required for its biodegradation involving the action of glucanases (endo and exo), cellobiohydrolase, and β-glucosidase.

10.4.2 Hemicellulose

Hemicellulose that occurs up to 20%–35% of the lignocellulosic material is a heteropolysaccharide composed of a variety of sugars as given in Fig. 10.4 (Limayem and Ricke, 2012). The hemicellulosic composition is not fixed in all sources instead shows variations from species to species. The branching points also differ in different sources generating more variation. One of the major components is xylan, which is a xylose polymer (Senthilkumar et al., 2005). The enzymes required for biodegradation of hemicellulose are endo-1,4-β-xylanase and xylan 1,4-β-xylosidase (Behnam et al., 2016).

10.4.3 Lignin

Lignin is the third component of lignocellulosic material and is present at a concentration of 15%–20% of the lingo cellulosic material. It is present more for structural plant support and is a natural sieve resistant toward the inflow of enzymes (Brodeur et al., 2011; Jogaiah et al., 2020). It is composed of polyphenols like sinapyl alcohol, p-coumaryl alcohol, and coniferyl alcohol (Agbor et al., 2011).

10.4.4 Starch

Starch, yet another major component of plant material, contains two main components—amylose and amylopectin. Amylose comprises linear polysaccharides

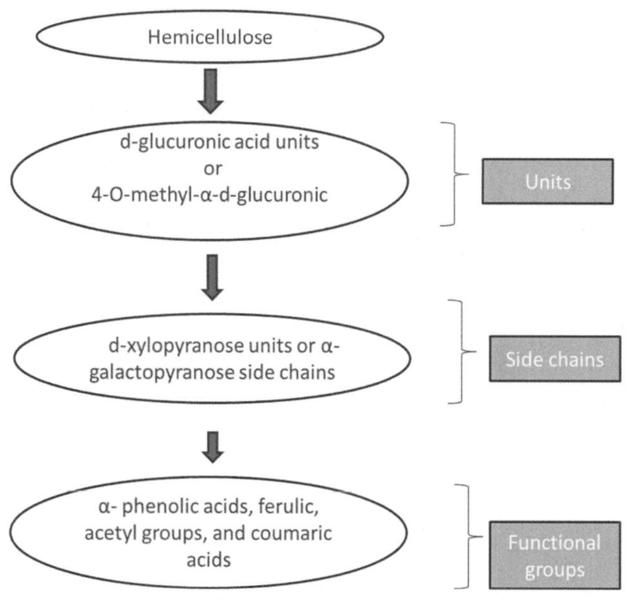

FIGURE 10.4

Major constituents of hemicellulose.

of glucose monomers linked by α-1,4-glycosidic bonds. The major component amylopectin contains a branched chain of glucose units, at branch points α-1,6-glycosidic bonds are encountered. The enzyme amylases efficiently hydrolyses the starch into short oligosaccharides or monomeric glucose (Homaei et al., 2016).

The disposal of this voluminous waste is a problem that is growing serious day by day (Rodríguez-Couto, 2008). The agroindustrial waste due to its composition and quantity offers an environmental threat if carelessly disposed. As they pose a problem to the human, animal, and environmental health, they are conventionally disposed by means like dumping or burning (Akhavan Sepahy et al., 2011; Geetha et al., 2021). The major deleterious environmental impact of the improper disposal procedures of agroindustrial is the generation of an enormous amount of greenhouse gases (Sadh et al., 2018). The waste generated by the juice industry or fruit industry is rich in nutrients whereas materials like husks are fiber-rich; therefore, nowadays instead of agroindustrial, they are acknowledged as raw material due to immense potential for the production of bioactive molecules. Presently, this chapter will focus on the utilization of agroindustrial residues exploited for the production of industrially important enzymes (Facchini et al., 2011). Microbial sources of enzymes are considered to be favorable due to their stability and easy reproducibility. Fungi, bacteria, and actinomycetes are the main sources of enzymes.

Bacillus species is gaining much industrial importance as it expresses many industrially important enzymes that cover nearly half of the industrial demand of value 1.6 billion dollars (Schallmey et al., 2004). Bacillus species is well known for its luxurious growth in simple laboratory conditions and easy culture maintenance with no specific nutritional need (Kunst and Rapoport, 1995). Generally, industries rely upon enzymes catalyzing catabolic reactions viz. hydrolytic reactions of enzymes such as cellulases, fructosyltransferase, lipase, proteases, and many more (Konsoula and Liakopoulou-Kyriakides, 2007).

10.5 Fructosyltransferase

This is one of the most important enzymes gaining popularity in the health market due to its ability to produce fructooligosaccharides that have wide applications as prebiotics and artificial sweeteners and encompass varied health benefits to the consumers. Attempts have been made by several research groups to economically produce these enzymes by exploiting cheap and easily available waste material. Table 10.1 describes the variety of wastes supporting the growth of microbes and the production of the enzyme.

10.5.1 Cellulases

Cellulase enzymes apart from conventional application in textile, pulp, paper, and detergent industries now are extended for biofuel production, in animal feed and

Table 10.1 Agroindustrial wastes used for fructosyltransferase production.

Microorganism	Agroindustrial waste	Enzyme produced
Aspergillus flavus NFCCI 2364 Ganaie et al. (2017)	Sugar cane bagasse	197.10 U/gds 423.18 U/gds (RCCD)
Aspergillus tamarii Kita Batista et al. (2020)	Wheat bran	1629.03 U/gds
Aspergillus carbonarius PC-4 Nascimento et al. (2019)	Pineapple crown	14.60 U/mL
Aspergillus niger GH1 Niger PSH *Penicillium citrinum* *Penicillium purpurogenum* Flores-Maltos et al. (2018)	Sugarcane bagasse Sotol bagasse Agave fibers polyurethane	2.70 g/g (Agave fibers using *A. niger* PSH)
Bacillus subtilis NCIM 2439 Sarat Babu et al. (2008)	Molasses	78.92 U/mL
Aspergillus niger Lateef et al. (2012)	Kola nut pod Ripe plantain pee	20.77 and 27.77 U/g
Penicillium citrinum Jayalakshmi et al. (2021)	Banana Banana peel Papaya Cane molasses	3.2 ± 0.5 6.9 ± 0.15 3.2 ± 0.56 7.3 ± 0.29

food industry (Sukumaran et al., 2005; Dakshayani et al., 2015). On the other hand, hemicellulases are used in the clarification of various industrial fluid products such as juices, mashes, bleaching agents, quality enhancement of feed, fodder, fibers, and for bioconversion of hemicelluloses to xylose sugars (Soni and Kango, 2013).

Cellulases have been reaped from microbes for long. Xia and Cen (1999) successfully exploited the industrial waste generated from the xylose industry for fermentative production of Cellulase using *Trichoderma reesei* ZU-02 as the catalyzing organism. The solid-state fermentation was carried out in tray fermentors and the highest productivity was up to 158 IFPU/g koji. 84% of saccharification was obtained after the scaling up process in trough fermentor.

As obvious, the process is simple, a suitable agroindustrial waste selection as a potential source for enzyme production and selection of potent microbe are critical features to be governed. The most striking quality of agroindustrial waste is the lignocellulosic content. Abdullah et al. (2016) successfully utilized municipal solid wastes for cellulase production using synergistic fermentation by *Trichoderma reesei* and *Aspergillus niger*. Optimization of cellulase production was also attempted for parameters like period of incubation, temperature, moisture, and inoculums size yielding enzyme productivity 26.10 ± 3.09 FPU/g (Abdullah et al., 2016).

The novel agroindustrial wastes used in cellulase production also produced considerable cellulase enzymes such as palm kernel cake, cow dung, digestate, and compost. The substrates such as banana, banana fruit stalk, banana peel, rice straw, rice husk, corn cob residue, and coconut coir are some of the commonly employed substrates for cellulase production. Some of the substrates from agroindustrial wastes utilized for cellulase production are given in Table 10.2. The common microbes expressing the cellulose enzyme at desirable levels are *Bacillus* sp., *Cellulomonas flavigena*, *Aspergillus* sp., *Trichoderma reesei*, and many more.

10.5.2 Proteases

Proteases are also gaining industrial popularity due to extended applications in various industrial processes and have bagged nearly 30% of the total industrial enzyme market demand (Schallmey et al., 2004). As discussed in earlier chapters, proteases find applications in diverse areas of life such as food, pharmaceuticals, laundry, and fabric (Nascimento et al., 2007). As stated in the introduction of this chapter, bacillus species is the most industrially favored enzyme source. Its main application is in the large-scale production of enzymes.

A novel approach of protease production and enhancement was attempted by Elumalai et al. (2020) using *Bacillus subtilis* B22 isolated from Kimchi. The enzyme production was conventional but aided with different light radiation availability using Led diodes. Blue, green, red, white, fluorescent, and dark conditions were provided to the bacteria during the growth and production period. This approach was employed using agroindustrial wastes like groundnut oil cake that offered satisfactory enzyme production.

Table 10.2 Utilization of agroindustrial waste for cellulase production.

Microorganism	Agroindustrial waste	References
Aspergillus sp.	Coir waste	Mrudula and Murugammal (2011)
	Wheat straw Wheat bran Rice straw Corn cob	Abo-State et al. (2010)
	Sugarcane bagasse Palm kernel cake	Lee et al. (2011)
	Corn husk Banana leaves Cotton seed meal	Kulkarni et al. (2018)
Bacillus subtilis MS 54	Maize bran Almond shell	Sharma and Bajaj (2017)
	Wheat bran	Kaur et al. (2018)
Bacillus sphaericus-JS1	Municipal solid wastes	Abdullah et al. (2016)
Cellulomonas genus	Rice straw and wastepaper	Sangkharak et al. (2011)
Trichoderma reesei RUT C30	Wheat bran	Singhania et al. (2007)
Trichoderma reesei	Water hyacinth	Deshpande and Shubhangi (2008)
	Rice bran Rice husk Rice straw	Darabzadeh and Parisa (2018)
	Sugarcane bagasse Palm kernel cake	Lee et al. (2011)
	Digestate and compost	Mejias et al. (2018)

Bacillus sp. cultured on brewery wastes has been experimented for coupled enzyme amylase and protease production as both are industrially important. The culture conditions were found analogous to the amylase production. The protease production was achieved at a higher pH where the enzyme production was analyzed metabolites linked to the growth curve (Blanco et al., 2016).

A very uncommon waste was utilized for protease production viz soy fiber, hair, and sludge mixture. The substrates were chosen for easy recovery of the process and low production cost. Various extraction strategies were screened for maximum enzyme recovery. This approach was also targeted to assess the residue left after enzyme production. By-products like biogas produced were analyzed in parallel to achieve a wasteless enzyme production process.

The volarization of soy fiber substrate was reported to be 91% and for hair sludge mixture was 121% respectively. The enzyme activity was 95 $\pm$ 6% and 94 $\pm$ 6% for soy fiber and hair sludge mixture respectively after downstream processing of the product after free drying and resuspension. The experiment also intended to lower the cost of production by using solvent ratio 1:3 *w*:*v* (Static, Soya fiber), and 1:2 *w*:*v* (agitated mode, Hair sludge mixture). The extraction was performed with

Table 10.3 Agroindustrial wastes used for protease production.

Microorganism	Agroindustrial wastes	References
Penicillium sp	Soya protein hydrolysate	Agrawal et al. (2004)
Aspergillus oryzae	Sesame oil cake	Ram and Kumar (2018)
Bacillus cereus	Corn cob Lentil husk	Prakash and Nigam (2013)
A. oryzae.	Coconut oil cake Rice bran Spent brewing grain Wheat bran Rice husk Palm kernel cake Sesame oil cake Olive oil cake Jackfruit seed	Sandhya et al. (2005)
Bacillus sp.	Wheat bran Red gram Green gram Black gram Chick pea Husks	Prakasham et al. (2006)
Aspergillus versicolor CJS-98	Deoiled Jatropha seed cake	Kumar et al. (2013)
Trichoderma viridiae strain VPG 12	Red gram husk Green gram husk Bengal gram husk	Vanegaon et al. (2014)
Bacillus sp.	Surface-modified coffee pulp Waste Corncobs	Kandasamy et al. (2016).
A. oryzae	Potato pulp powder	Murthy and Kusumoto (2015)
Pseudomonas aeruginosa	Wheat bran	Meena et al. (2013)

distilled water. The complete waste treatment is suggested by composting and anaerobic sludge digestion (Marin et al., 2018). Table 10.3 insights the variety of wastes experimented with for protease production.

10.6 Xylanase

Xylanases are clarifying enzymes finding applications in the juice industry, food, feed, and fodder. Fungal and bacterial xylanases were produced using substrates as wheat bran, sugarcane bagasse, rice straw, and soya bean hulls, rice bran, grape pomace, orange peel, rice husk (Gawande and Kamat, 1999; Soppina et al., 2005; Botella et al., 2007; Patel and Prajapati, 2014).

The production of xylanase supplemented with wheat bran and anaerobically treated distillery spent wash by *Aspergillus foetidus* MTCC 4898 was successfully attempted using response surface methodology involving Box–Behnken design (BBD) xylanase optimization. The optimization of factors was modeled involving pH, moisture content, and percentage of inoculum. The model exhibited xylanase production 8200–8400 U/g. The experimental value reported was 8450 U/g. The agroindustrial waste was pretreated with dilute NaOH and ammonia. The saccharification of wheat straw, rice straw, and corncobs were investigated yielding 4.9, 4.7, 4.6 g/L reducing sugars, respectively (Chapla et al., 2010).

A similar approach of using anaerobically treated distillery spent wash complimented with rice straw was attempted for cellulase and xylanase production. The BBD was selected for the maximization of enzyme production using *Aspergillus heteromorphus*. The factors chosen for interaction and their influence on enzyme production were pH, temperature, spent wash, and rice straw. The design suggested the optimized parameters to be 6% and 3% of spent ash and rice straw, respectively, at pH 5 and temperature 32.5°C. This study suggested a novel combination of wastes for enzyme production (Bajar et al., 2020). Experiments based on a variety of agricultural wastes for xylanase enzyme have been attempted innumerable times. Table 10.4 elaborates the microbes used for the production of xylanase from various agroindustrial wastes.

Table 10.4 Agroindustrial wastes utilized for xylanase production.

Microorganism	Agroindustrial waste	References
Aspergillus niger KIBGE-IB3	Corncob Wheat bran Rice husk Orange peel Pomegranate peel	Javed et al. (2019)
A. terreus and *A. niger*	Wheat bran Sugarcane bagasse Rice straw Soya bean hulls	Gawande and Kamat (1999)
Aspergillus awamori.	Grape pomace	Botella et al. (2007)
Streptomyces	Sugarcane bagasse Wheat bran Rice bran Corn cob Wheat straw	Singh et al. (2012)
B. subtilis ATCC 6633	Barley husk	Ho and Chinonso (2016)
Chaetomium globosum	Pomegranate peel	Atalla and El Gamal (2020)
Paecilomyces themophila J18	Wheat straw	Yang et al. (2006)
Streptomyces actuosus A-151	Rice bran	Wang et al. (2003)
Streptomyces thermocarboxydus	Wheat bran	Tran et al. (2021)
Thermomyces lanuginosus	Sorghum straw	Sonia et al. (2005)

10.7 Conclusion

The percentage of industries demanding enzymes are now expanding as the enzyme is coming up with a robust nature and decreased sensitivity. The market trend clearly indicates the steep increase in enzyme demand not only in quantity but in different types desired for versatile processes involved in sectors of food, feed, pharmaceuticals, etc. The industry that values enzymes most is the food sector in which enzyme is an unparallel component of many processes, indispensable in the food industry. The enzymes discussed in this chapter and many more are of great commercial value offering an environmentally friendly path for agroindustrial wastes to be recycled. The fermentative use of agroindustrial wastes for the generation of biocatalytic molecules is to be scaled up for industries to opt for this process. Although the resistant nature of the lignocellulose material offers a hindrance in direct bioconversion of energy trapped in organic material to useable form yet green technology can be used for preliminary treatment of immensely resistant agroindustrial wastes. It would be preferred if the term waste can be replaced by agroindustrial substrates for enzyme production. The agroindustrial-based enzyme production can be scaled up using the advancement in research and development. More new technologies should be designed to make the cost of enzyme production low. The example of Novozymes is inspiring for other industries in the enzyme market that utilizes process based on soy grits and corn starch for enzyme production. Industries should be promoted to opt for fermentation processed based on agroindustrial substrates for enzyme production.

References

Abdullah, J.J., Greetham, D., Pensupa, N., Tucker, G.A., Du, C., 2016. Optimizing cellulase production from municipal solid waste (MSW) using solid state fermentation (SSF) renewable. Energy Appl. J. Fundam. 6, 3. https://doi.org/10.4172/2090-4541.1000206.

Abo-State, M.A.M., Hammad, A.I., Swelim, M., Gannam, R.B., 2010. Enhanced production of cellulase(S) by *Aspergillus* spp. isolated from agriculture wastes by solid state fermentation. Am. Euras. J. Agric. Environ. Sci. 8 (4), 402–410.

Agamuthu, P., 2009. Challenges and opportunitics in Agrowaste management: an Asian perspective. In: Inaugural meeting of First Regional 3R Forum in Asia 11–12 Nov., Tokyo, Japan.

Agbor, V.B., Cicek, N., Sparling, R., Berlin, A., Levin, D.B., 2011. Biomass pretreatment: fundamentals toward application. Biotechnol. Adv. 29 (6), 675–685.

Agrawal, D., Patidar, P., Banerjee, T., Patil, S., 2004. Production of alkaline protease by Penicillium sp. under SSF conditions and its application to soy protein hydrolysis. Process Biochem. 39, 977–998.

Akhavan Sepahy, A., Ghazi, S., Akhavan Sepahy, M., 2011. Cost-effective production and optimization of alkaline xylanase by indigenous *Bacillus mojavensis* AG137 fermented on agricultural waste. Enzym. Res. https://doi.org/10.4061/2011/593624, 593624.

Atalla, S.M.M., El Gamal, N.G., 2020. Production and characterization of xylanase from pomegranate peel by *Chaetomium globosum* and its application on bean under greenhouse condition. Bull. Natl. Res. Cent. 44, 104. https://doi.org/10.1186/s42269-020-00361-5.

Babu, A.N., Jogaiah, S., Ito, S.I., Amruthesh, K.N., Tran, L.S.P., 2015. Improvement of growth, fruit weight and early blight disease protection of tomato plants by rhizosphere bacteria is correlated with their beneficial traits and induced biosynthesis of antioxidant peroxidase and polyphenol oxidase. Plant Sci. 231, 62–73. https://doi.org/10.1016/j.plantsci.2014.11.006.

Bajar, S., Singh, A., Bishnoi, N.R., 2020. Exploration of low-cost agro-industrial waste substrate for cellulase and xylanase production using *Aspergillus heteromorphus*. Appl. Water Sci. 10, 153. https://doi.org/10.1007/s13201-020-01236-w.

Batista, J.M.S., Romero, M.P., Brandão-Costa, Márcia, N., Cunha, C.D., Hélio, O.S., Rodrigues, Porto, L.F., 2020. Purification and biochemical characterization of an extracellular fructosyltransferase-rich extract produced by *Aspergillus tamarii* Kita UCP1279. Biocatal. Agric. Biotechnol. 26, 101647.

Behnam, S., Karimi, K., Khanahmadi, M., Salimian, Z., 2016. Optimization of glucoamylase production by *Mucor indicus, Mucor hiemalis*, and *Rhizopus oryzae* through solid state fermentation/*Mucor indicus, Mucor hiemalis, Rhizopus oryzae* tarafından üretien glukoamilazın katı hal fermantasyonu ile optimizasyonu. Turkish J. Biochem. 41. https://doi.org/10.1515/tjb-2016-0036.

Bhavya, G., Belorkar, S.A., Mythili, R., Geetha, N., Shetty, H.S., Udikeri, S.S., Jogaiah, S., 2021. Remediation of emerging environmental pollutants: a review based on advances in the uses of eco-friendly biofabricated nanomaterials. Chemosphere 275, 129975.

Blanco, A.S., Durive, O.P., Pérez, S.B., Montes, Z.D., Guerra, N.P., 2016. Simultaneous production of amylases and proteases by Bacillus subtilis in brewery wastes. Braz. J. Microbiol. 47, 665–674.

Botella, C., Diaz, A., Ory, I., Webb, C., Blandino, A., 2007. Xylanase and pectinase production by Aspergillus awamori on grape pomace in solid state fermentation. Process Biochem. 42, 98–101.

Brodeur, G., Yau, E., Badal, K., Collier, J., Ramachandran, K.B., Ramakrishnan, S., 2011. Chemical and Physicochemical Pretreatment of Lignocellulosic Biomass: A Review, Enzyme Research. https://doi.org/10.4061/2011/787532. Article ID 787532.

Chapla, D.D., Jyoti, M.D., Shah, A., 2010. Utilization of agro-industrial waste for xylanase production by *Aspergillus foetidus* MTCC 4898 under solid state fermentation and its application in saccharification. Biochem. Eng. J. 49, 361–369. https://doi.org/10.1016/j.bej.2010.01.012.

Dakshayani, S.D.S., Sharathchandra, R.G., Jogaiah, S., 2015. Screening of endophytic fungi for their ability to produce extracellular cellulases. Int. J. Pharm. Pharmaceut. Sci. 7, 205–211.

Darabzadeh, N.H.E., Parisa, Z.H., 2018. Optimization of cellulase production under solid-state fermentation by a new mutant strain of *Trichoderma reesei*. Food Sci. Nutr. 7. https://doi.org/10.1002/fsn3.852.

Deshpande, P.N., Shubhangi, K.S., 2008. Water hyacinth as carbon source for the production of cellulase by *Trichoderma reesei*. Appl. Biochem. Biotech. 158, 552–560. https://doi.org/10.1007/s12010-008-8476-9.

Elumalai, P., Lim, J.M., Park, Y.J., et al., 2020. Agricultural waste materials enhance protease production by *Bacillus subtilis* B22 in submerged fermentation under blue light-emitting diodes. Bioproc. Biosyst. Eng. 43, 821–830. https://doi.org/10.1007/s00449-019-02277-5.

Facchini, F.D.A., Vici, A.C., Reis, V.R.A., Jorge, J.A., Terenzi, H.F., Reis, R.A., Polizeli, M., de LT de, M., 2011. Production of fibrolytic enzymes by *Aspergillus japonicus* C03 using agro-industrial residues with potential application as additives in animal feed. Bioproc. Biosyst. Eng. 34, 347–355. https://doi.org/10.1007/s00449-010-0477-8.

Fang, J., Jogaiah, S., Guan, L., Sun, X., Abdelrahman, M., 2018. Coloring biology in grape skin: a prospective strategy for molecular farming. Physiol. Plantarum 164, 429–441. https://doi.org/10.1111/ppl.12822.

Flores-Maltos, A., Mussatto, S.I., Contreras-Esquivel, J., Rodríguez, R., Teixeira, J., Aguilar, C.N., 2018. Production of a transfructosylating enzymatic activity associated to fructooligosaccharides. In: Parameswaran, B., Varjani, S., Raveendran, S. (Eds.), Green Bio-Processes. Springer, pp. 345–355. https://doi.org/10.1007/978-981-13-3263-0_18.

Ganaie, M.A., Soni, H., Naikoo, G.A., Oliveira, L.T.S., Rawat, H.K., Mehta, P.K., Narain, N., 2017. Screening of low cost agricultural wastes to maximize the fructosyltransferase production and its applicability in generation of fructooligosaccharides by solid state fermentation. Int. Biodeterior. Biodegrad. 118, 19–26.

Gawande, P.V., Kamat, M.Y., 1999. Production of *Aspergillus* Xylanase by Lignocellulosic Waste Fermentation and its Application, vol. 87, pp. 511–519.

Geetha, N., Bhavya, G., Abhijith, P., Shekhar, R., Dayananda, K., Jogaiah, S., 2021. Insights into nanomycoremediation: secretomics and mycogenic biopolymer nanocomposites for heavy metal detoxification. J. Hazard Mater. 409, 124541. https://doi.org/10.1016/j.jhazmat.2020.124541.

Haider, M.S., Khan, N., Pervaiz, T., Zhongjie, L., Jogaiah, S., Jiu, S., Fang, J., 2019. Genome-wide identification, evolution, and molecular characterization of the PP2C gene family in Woodland Strawberry. Gene 702, 27–35. https://doi.org/10.1016/j.gene.2019.03.025.

Ho, H., Chinonso, A., 2016. Overproduction of xylanase from mutants of *Bacillus subtilis* with barley husk as the prime carbon source under submerged fermentation after random mutagenesis using ethyl methane sulfonate (EMS) and acridine orange (AO). Br. Microbiol. Res. J. 14, 1–17. https://doi.org/10.9734/BMRJ/2016/22959.

Homaei, A., Ghanbarzadeh, M., Monsef, F., 2016. Biochemical features and kinetic properties of α-amylases from marine organisms. Int. J. Biol. Macromol. 83, 306–314.

Javed, U., Ansari, A., Aman, A., Qader, S.L.U., 2019. Fermentation and saccharification of agro-industrial wastes: a cost-effective approach for dual use of plant biomass wastes for xylose production. Biocatal. Agric. Biotechnol. 21, 101341.

Jayalakshmi, J., Mohamed Sadiqa, A., Sivakumar, V., 2021. Microbial enzymatic production of fructooligosaccharides from sucrose in agricultural harvest. Asian J. Microbiol. Biotech. Environ. Sci. 23, 84–88.

Jogaiah, S., Praveen, S., De Britto, S., Konappa, N., Udayashankar, A.C., 2020. Exogenous priming of chitosan induces upregulation of phytohormones and resistance against cucumber powdery mildew disease is correlated with localized biosynthesis of defense enzymes. Int. J. Biol. Macromol. 162, 1825–1838. https://doi.org/10.1016/j.ijbiomac.2020.08.124.

Kandasamy, S., Muthusamy, G., Balakrishnan, S., Duraisamy, S., Thangasamy, S., Seralathan, K.K., Chinnappan, S., 2016. Optimization of protease production from surface-modified coffee pulp waste and corncobs using *Bacillus* sp. SSF. 3 Biotech 6, 167. https://doi.org/10.1007/s13205-016-0481-z.

Kaur, P., Bhardwaj, S., Bhardwaj, N.K., Sharma, J., 2018. Lignocellulosic waste as a sole substrate for production of crude cellulose from *Bacillus subtilis* PJK6 under solid state fermentation using statistical approach. J. Carbohydra. 1, 1–14.

Konsoula, Z., Liakopoulou-Kyriakides, M., 2007. Co-production of alpha-amylase and beta-galactosidase by *Bacillus subtilis* in complex organic substrates. Bioresour. Tech. 98, 150–157.

Kulkarni, N., Vaidya, T., Rathi, G., 2018. Optimization of cellulase production by Aspergillus species under solid state fermentation. Pharm. Innov. J. 7 (1), 193–196.

Kumar, M., Shivakumar, M.V.S., Somashekar, S.D., 2013. Solid-state fermentation of Jatropha seed cake for optimization of lipase, protease and detoxification of anti-nutrients in Jatropha seed cake using Aspergillus versicolor CJS-98. J. Biosci. Bioeng. 117. https://doi.org/10.1016/j.jbiosc.2013.07.003.

Kunst, F., Rapoport, G., 1995. Salt stress is an environmental signal affecting degradative enzyme synthesis in *Bacillus subtilis*. J. Bacteriol. 177, 2403–2407.

Lateef, A., Oloke, J., Gueguim-Kana, E.B., Raimi, O.R., 2012. Production of fructosyltransferase by a local isolate of *Aspergillus niger* in both submerged and solid substrate media. Acta Aliment. 40, 100–117. https://doi.org/10.1556/AAlim.41.2012.1.12.

Lee, C.K., Darah, I., Ibrahim, C.O., 2011. Production and optimization of cellulase enzyme using *Aspergillus niger* USM AI 1 and Comparison with *Trichoderma reesei* via solid state fermentation system. Biotechnol. Res. Int. 2011. https://doi.org/10.4061/2011/658493. Article ID 658493.

Limayem, A., Ricke, S., 2012. Lignocellulosic biomass for bioethanol production: current perspectives, potential issues and future prospects. Prog. Energy Combust. Sci. 38, 449–467. https://doi.org/10.1016/j.pecs.2012.03.002.

Marín, M., Artola, A., Sánchez, A., 2018. Production of proteases from organic wastes by solid-state fermentation: downstream and zero waste strategies. 3 Biotech 8 (4), 205. https://doi.org/10.1007/s13205-018-1226-y.

Meena, P., Tripathi, A.D., Srivastava, S.K., Jha, A., 2013. Utilization of agro-industrial waste (wheat bran) for alkaline protease production by *Pseudomonas aeruginosa* in SSF using Taguchi (DOE) methodology. Biocatal. Agric. Biotechnol. 2, 210–216.

Mejias, L., Cerda, A., Barrena, R., Gea, T., Sánchez, A., 2018. Microbial strategies for cellulase and xylanase production through solid-state fermentation of digestate from biowaste. Sustainability 10, 2433. https://doi.org/10.3390/su10072433.

Mrudula, S., Murugammal, R., 2011. Production of cellulase by Aspergillus Niger under submerged and solid state fermentation using coir waste as a substrate. Braz. J. Microbiol. 42, 1119–1127. https://doi.org/10.1590/S1517-83822011000300033.

Murthy, P.S., Kusumoto, K.I., 2015. Acid protease production by *Aspergillus oryzae* on potato pulp powder with emphasis on glycine releasing activity: a benefit to the food industry. Food Bioprod. Process. 96, 180–188.

Nascimento, W.C.A., Silva, C.R., Carvalho, R.V., Martins, M.L.L., 2007. Otimização de um meio de cultura para a produção de proteases por um *Bacillus* sp. Ciência e Tecnologia de Alimentos. 27, 417–421.

Nascimento, G.C.do, Batista, R.D., Amaral Santos, C.C.A.do, da Silva, E.M., Coutinho de Paula, F., Mendes, D.B., Paula de Oliveira, D., Fernando de Almeida, A., 2019. β-Fructofuranosidase and β –D-Fructosyltransferase from new *Aspergillus* carbonarius PC-4 strain isolated from canned peach syrup: effect of carbon and nitrogen sources on enzyme production. Sci. World J. 2019. https://doi.org/10.1155/2019/6956202. Article ID 6956202.

Obi, F.O., Ugwuishiwu, B.O., Nwakaire, J.N., 2016. Agricultural waste concept, generation, utilization and management Nigerian. J. Technol. 35, 957–964.

Pappu, A., Saxena, M., Asolekar, S.R., 2007. Solid wastes generation in India and their recycling potential in building materials. Build. Environ. 42, 2311–2320.

Patel, K., Kalavati, P., 2014. Xylanase production by *Cladosporium* sp. from agricultural waste. Int. J. Curr. Res. Acad. Rev. 2, 84–90.

Patil, S.V., Kumudini, B.S., Pushpalatha, H.G., De Britto, S., Ito, S-ichi, Sudheer, S., Singh, D.P., Gupta, V.K., Jogaiah, S., 2020. Synchronised regulation of disease resistance in primed finger millet plants against the blast disease. Biotechnol. Rep. 27, e00484. https://doi.org/10.1016/j.btre.2020.e00484.

Prakash, M., Nigam, S., 2013. Cost Effective Method of Protease Production in Solid State Fermentation Using Combined Substrate Corn Cob and Lentil Husk, vol. 3, pp. 2249−8656.

Prakasham, R.S., Rao, C.S., Sarma, P.N., 2006. Green gram husk—an inexpensive substrate for alkaline protease production by *Bacillus* sp. Solid-State Ferment. Bioresour. Tech. 97, 1449−1454.

Ram, M.R., Kumar, S., 2018. Production of alkaline protease from *Aspergillus oryzae* isolated from seashore of Bay of Bengal. J. Appl. Nat. Sci. 10 (4), 1210−1215.

Rodríguez Couto, S., 2008. Exploitation of biological wastes for the production of value-added products under solid-state fermentation conditions. A Sour. Bioact. Comp. 3, 859−870.

Sadh, P.K., Duhan, S., Duhan, J.S., 2018. Agro-industrial wastes and their utilization using solid state fermentation: a review. Bioresour. Bioproc. 5, 1. https://doi.org/10.1186/s40643-017-0187-z.

Sandhya, C.A., George, S.S., Pandey, A., 2005. Comparative evaluation of neutral protease production by *Aspergillus oryzae* in submerged and solid-state fermentation. Process Biochem. 40, 2689−2694. https://doi.org/10.1016/j.procbio.2004.12.001.

Sangeetha, J., Thangadurai, D., Hospet, R., Purushotham, P., Manowade, K.R., Jogaiah, S., 2017. Production of bionanomaterials from agricultural wastes. Nanotechnology 33−58.

Sangkharak, K., Vangsirikul, P., Janthachat, S., 2011. Isolation of novel cellulase from agricultural soil and application for ethanol production. Int. J. Adv. Biotechnol. Res. 2, 230−239.

Sarat Babu, I., Ramappa, S., Guru Mahesh, D., Sunanda Kumari, K., Sita Kumari, K., Subba, G., 2008. Optimization of medium constituents for the production of fructosyltransferase (ftase) by *Bacillus subtilis* using Response surface methodology. Res. J. Microbiol. 3, 114−121.

Schallmey, M., Singh, A., Ward, O.P., 2004. Developments in the use of Bacillus species for industrial production. Can. J. Microbiol. 50 (1), 1−17. https://doi.org/10.1139/w03-076.

Senthilkumar, S., Ashokkumar, B., Chandraraj, K., Gunasekaran, P., 2005. Optimization of medium composition for alkali-stable xylanase production by Fxn 1 in solid-state fermentation using central composite rotary design. Bioresour. Technol. 96, 1380−1386. https://doi.org/10.1016/j.biortech.2004.11.005.

Sharma, M., Bajaj, B.K., 2017. Optimization of bioprocess variables for production of a thermostable and wide range pH stable carboxymethyl cellulase from *Bacillus subtilis* MS 54 under solid state fermentation. Environ. Prog. Sustain. Energy 36, 1123−1130.

Singh, R., Kapoor, V., Kumar, V., 2012. Utilization of agro-industrial wastes for the simultaneous production of amylase and xylanase by thermophilic Actinomycetes. Braz. J. Microbiol. 43, 1545−1552.

Singhania, R.R., Sukumaran, R.K., Pandey, A., 2007. Improved cellulase production by *Trichoderma reesei* RUT C30 under SSF through process optimization. Appl. Biochem. Biotechnol. 142, 60−70. https://doi.org/10.1007/s12010-007-0019-2.

Soni, H., Kango, N., 2013. Microbial mannanases: properties and applications. In: Shukla, P., Pletschke, B.I. (Eds.), Advances in Enzyme Biotechnology. Springer, New York, pp. 41−56.

Sonia, K.G., Chadha, B.S., Saini, H.S., 2005. Sorghum straw for xylanase hyper-production by *Thermomyces lanuginosus* (D2W3) under solid-state fermentation. Bioresour. Technol. 96 (14), 1561–1569. https://doi.org/10.1016/j.biortech.2004.12.037.

Soppina, V.B., Gaikwad, G., Gaikwad, S.R., Naik, G.R., 2005. Production of a xylanolytic enzyme by a thermoalkaliphilic *Bacillus* sp. JB-99 in solid state fermentation. Process Biochem. 40, 431–435. https://doi.org/10.1016/j.procbio.2004.01.027.

Sudisha, J., Amruthesh, K.N., Deepak, S.A., Shetty, N.P., Sarosh, B.R., Shekar Shetty, H., 2005. Comparative efficacy of strobilurin fungicides against downy mildew disease of pearl millet. Pest. Biochem. Physiol. 81, 188–197. https://doi.org/10.1016/j.pestbp.2004.08.001.

Sukumaran, R.K., Singhania, R.R., Pandey, A., 2005. Microbial cellulases—production, applications and challenges. J. Sci. Ind. Res. 64, 832–844.

Tran, T.N., Doan, C.T., Wang, S.L., 2021. Conversion of wheat bran to xylanases and dye adsorbent by *Streptomyces thermocarboxydus*. Polymers 13 (2), 287. https://doi.org/10.3390/polym13020287.

Vanegaon, S.K., Hanchinalmath, J., Sundeep, Y., Borah, D., Talluri, V.P., 2014. Optimization and production of alkaline proteases from agro byproducts using a novel trichoderma viridiae strain VPG 12, isolated from agro soil. Int. Lett. Nat. Sci. 9, 78–84. https://doi.org/10.18052/www.scipress.com/ILNS.14.77.

Wang, S.L., Yen, Y.H., Shih, I.L., Chang, A.C., Wu, W.T., Chai, W.C., Yue, D., 2003. Production of xylanases from rice bran by Streptomyces actuosus A-151. Enzym. Microb. Technol. 33, 917–925. https://doi.org/10.1016/S0141-0229(03)00246-1.

Xia, L., Cen, P., 1999. Cellulase production by solid state fermentation on lignocellulosic waste from the xylose industry. Process Biochem. 34, 909–912.

Yang, S.Q., Yan, Q.J., Jiang, Z.Q., Li, L.T., Tian, H.M., Wang, Y.Z., 2006. High-level of xylanase production by the thermophilic *Paecilomyces themophila* J18 on wheat straw in solid-state fermentation. Bioresour. Tech. 97, 1794–1800.

Yu, J., Wang, D., Geetha, N., Khawar, M., Mujtaba, M., Jogaiah, S., 2021. Current trends and challenges in the synthesis and applications of chitosan-based nanocomposites for plants. Carbohydr. Polym. 261, 117904. https://doi.org/10.1016/j.carbpol.2021.117904.

CHAPTER

Significance of enzymes and their agricultural applications

11

11.1 Introduction

Enzymes that cater the potential of enhancing soil fertility, inhibit pests growth, and control plant pathogens are extraordinarily useful in agriculture (Satapute et al., 2019). Such enzymes are termed as agricultural enzymes. These enzymes are also proteins with specific catalytic or biotransformation feature as other industrial enzymes. These enzymes are gaining popularity in the agriculture sector as they are indispensable alternatives for chemicals used as pesticides, fertilizers, and antimicrobial agents (Dakshayani et al., 2015; Konappa et al., 2021).

In comparison to chemicals, the agricultural enzymes are eco-friendly, exhibit efficient crop improvement, and promote desirable growth for satisfactory enzyme yield. Thus, protection, improvement, and production are the three areas where agricultural enzymes are developing a potential market.

The growth of the agriculture enzyme market has crossed over USD 200 million since the last 5 years. The demand is consistently growing with the population rise and development of a section of consumers demanding more of organic food due to their health benefits. Looking at the potential rise in the agriculture enzyme market, the industries are now in search of technologies and strategies for the production of such agricultural enzymes at an economic level. At the international level, the market projects a Compound annual growth rate (CAGR) of 11.7%. As consumer awareness regarding food quality and health is increasing, the demand for organic food is experiencing exponential growth. The strict regulations of food and drug administration, the United States has given prime importance to the consistent performance of such biological products. If there is a slight deviation of the product, it is withdrawn from the market.

Table 11.1 cites major enzymes playing key roles in soil fertility improvement viz. phosphatases, dehydrogenases, and sulfatases (Babu et al., 2017). The major enzymes used in agriculture are enlisted in Table 11.1.

Presently, in agriculture, the use of whole organisms as biofertilizers is more prevalent. The use of enzymes such as phosphatases, dehydrogenases, and sulfatases are highlighted as a means for successful soil remediation. Phosphatases share the

Protocols and Applications in Enzymology. https://doi.org/10.1016/B978-0-323-91268-6.00008-9

Table 11.1 Major enzymes remediating soil fertility with their specified role.

Enzymes and their role in soil improvement	Ref.
Phosphatases	
Control the biotic pathways of phosphorus (P) Phosphatase activity associated with climatic conditions and soil type Channelizes the phosphorus cycle (P cycle), which correlates to plant growth and P stress (phosphorus stress) Phosphatase is an indicator of fertility, and it also indicates the phosphorus deficiency in the soil.	Nannipieri et al. (2015) Margalef et al. (2017)
Dehydrogenases	
Help in biological oxidation of soil Indicators of microbiological activities in soil Bioindicators for the removal of hydrocarbons from soil Determines the overall activity of microorganisms, especially in activated sludge or soil	Małachowska-Jutsz and Matyja (2019) Kumar et al. (2013); Järvan et al. (2014)
β-Glucosidase	
It is involved in catalyzing the hydrolysis of carbohydrates Biodegradation of other beta-glucosidase in plant debris Used for indicating soil quality and is related to the carbon cycle Acting in the cleavage of cellobiose into glucose molecules It is a sensitive enzyme and considered as a soil quality indicator It is directly related to the quantity and quality of soil organic matter. Lower enzyme levels indicate organic material with a high C:N ratio and high amounts of lignified roots Enzyme activity tends to be higher in soils with high content of easily decomposable organic matter It is also high in preserved soils and also in soils under crop rotation and direct planting The addition of the soil organic residues such as biosolids, manure, urban sludge, and poultry litter, also increases the activity of this enzyme in the soil	Ferraz de Almeida et al. (2015)
Urease	
Urease is used for the hydrolysis of urea fertilizers The products are NH3 and CO_2 It also increases the pH It is available as extra- and intracellular enzymes of both plants and microorganisms Urease is degraded in soil due to the action of proteolytic enzymes It is the indicator of the occurrence of accessibility of N for plant growth in the N cycle and in the widespread usage of urea as a fertilizer	Burak (2020)

biggest demand among all enzymes applicable in the improvement of soil fertility. In the market, the CAGR of phosphatase is expected to rise by 12.6% in the year 2027.

The agricultural enzymes are produced conventionally as any other industrially important enzyme by using a potent microbial source (Jogaiah et al., 2018). The fermentation medium is selected on the basis of the type of enzyme to be produced.

Crop production is deeply related and dependent on agriculture enzymes (Pushpalatha et al., 2013). The microbicidal enzymes help to reduce grain damage and increases the shelf life. Organic crop production is favored now by industries for a sector of consumers who are health conscious. The organic farming yields vegetables, fruits, and cereals grown free of any chemicals used as fertilizers or pesticides. This makes the food material very healthy, safe, nutritious, and free of side effects after consumption. Such food materials produced under the organic farming banner are enjoying popularity and increasing demand in the food and health market. Such food not only confers nutrition but added advantages of improved immunity, anticancerous, antiinflammatory, etc. These additional benefits have also influenced the demand hike of such enzymes in the enzyme market.

11.2 Role of enzymes in soil

Agriculture sector is solely dependent on the soil that is used for crop production. Soil health is vital for the crop and a critical factor that governs the yield. Soil caters microbes that are natural inhabitants and provide a great activity support for recycling the nutrients and maintaining the ecological balance. The study of enzymes in soil plays a significant role in predicting soil status or ill health. They can act as indicators of ecological balance involving organic matter decomposition and nutrient recycling (Piotrowska-Długosz, 2019).

11.2.1 Alkaline phosphatase

The phosphatases are one of the most important agricultural enzymes. This enzyme catalyzes the mobilization of the organic form of phosphorus into a useable form. Thus, this enzyme treatment initiates the release of phosphorus and increases its recycling through improved plant assimilation (Nalini et al., 2015) (Table 11.2).

The growth of plants is mainly dependent on nitrogen, phosphorus, and potassium available in the soil. Phosphorous plays a vital role in the synthesis of nucleic acids, as a building block of cell infrastructure and for cellular energy transaction and storage. The role of phosphatase in the soil is given in Fig. 11.1.

The phosphatase enzymes help in solubilization of phosphorus that can be assimilated by plants and microbes. The phosphatase enzyme available in soil depends upon the requirement of phosphorus by plants and microbes and the free available phosphorus in soil. The rhizosphere zone also exhibits phosphatase activity due to secretions associated with roots and mycorrhiza. The stability of phosphatase is dependent in turn on the surrounding soil, aluminum oxide, and also iron. Many

Table 11.2 Alkaline phosphatase as an indicator of soil health.

Soil characteristics	Ref.
Soil tillage and agricultural residues affect the activity of the enzyme	Deng and Tabatabai (1997)
Non till treatment of soils has increased enzyme activity	Wang et al. (2011)
Soil samples of acid farm fields and alkaline farm fields exhibit enzyme activity	Eivazi and Tabatabai (1977)
Levels of microbial alkaline phosphatase act as an indicator to determine the agricultural soil quality	Nalini et al. (2015)
The alkaline phosphatase activity acts as a pH adjustment indicator in determining the optimum soil pH required for crop production	Margalef et al. (2017)
Helps in measurements of the impact of genetically modified microorganisms in the ecosystem	Sharma et al. (2013)
Levels of the enzyme can be used to access the impact of transgenic plants	Zeng et al. (2016)
Indicates symbiotic efficiency of microbe in proximity of the roots	Maseko and Dakora (2013)

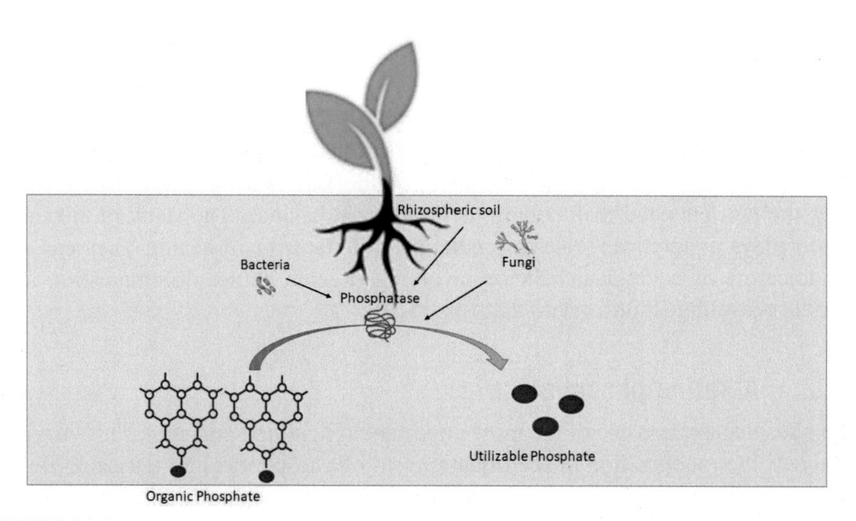

FIGURE 11.1

The source of phosphatase and its role in soil.

times, enzyme-active zones are encountered in soil without the presence of phosphatase-producing organisms.

11.2.2 Enzymes as biopesticides

According to the report of Globe news fire in June 2020. The global biopesticides market size is projected to grow at a CAGR of 14.7% from an estimated billion value of USD 4.3 billion in 2020 to reach USD 8.5 billion by 2025. The market has been gaining wide importance among farmers to produce residue-free food products with the usage of microbial-based bioinsecticides. Biological solutions have proved to be an effective alternative to conventional chemicals and even work optimally when applied in combination.

The principle used in enzymes as effective biopesticides is to target the chitin that is the building material of the exoskeleton and gut of insects. Chitinase is the enzyme that specifically degrades these polysaccharides to small carbohydrate chains. After rigorous experiments, chitinase has been successfully used as an effective biopesticide against insects and fungal pests. The transgenic plants have successfully shown resistance against insect pests that have incorporated the Chitinase gene from tobacco hornworm. Fig. 11.2 explains the application of chitinase in producing pest-resistant plant species (Kramer and Muthukrishnan, 1997).

The pesticides target sensitive systems in the insects to offer their pesticidal action. The major systems targeted by various pesticides are the nervous system, endocrine system, exoskeleton development, gut lining, and many more.

Yet another novel approach is to use the enzymes Juvenile hormone esterases and Juvenile hormone epoxide hydrolases. The metabolites produced by these enzymes are critically important in insect lifecycle. Approaches that are successful in inhibiting such enzymes would be promising for pest control. Recombinant DNA technology or chemicals are used as effective inhibitors of these enzymes (El-Sheikh et al., 2011; Tan et al., 2005; Hanzlik and Hammock, 1988).

11.2.3 Enzymes as antimicrobial agents

Many plant diseases are borne by pathogens infecting seeds. Such pathogens are harbored by seeds since their sowing leads to developmental hindrances like loss of germination capacity, necrosis, rot, or even abortion of seed. If the seed is successful in germinating, the accompanying pathogen reflects its damage in the later stage

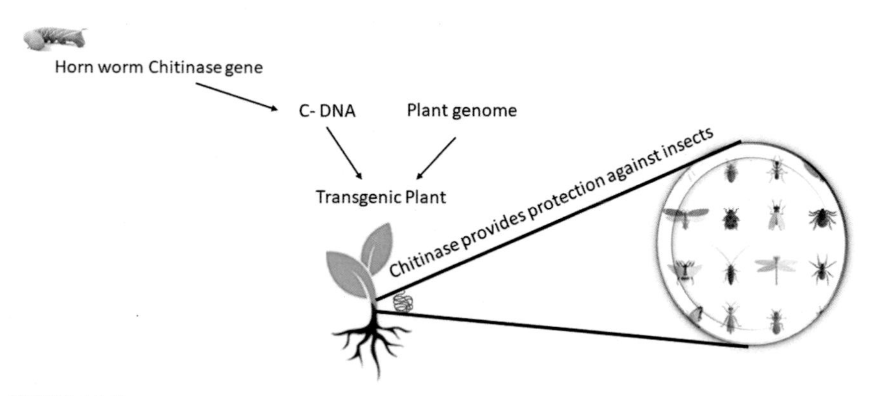

FIGURE 11.2

Use of chitinase for production of insect-resistant plants.

of plant growth in form of disease or plant infection (Sudisha et al., 2006; Singh et al., 2017).

Effective use of the antifungal property of the xylanase enzyme has been applied for seed treatment to make it pathogen-free. The crude xylanase successfully reduced the growth of *A. alternata*, *F. oxysporum*, *Phoma destructive*, *R. solani*, and *Sclerotium rolfsii*. Xylanase is good biocontrol for green and blue mold of orange and lime fruit (Atalla and El Gamal, 2020)

11.3 Conclusion

The soil health and its related functionalities are an indirect effect of enzymatic activities perpetuating in soil whether in free form or cell-bound form. As already overviewed, urease, on one hand, improvises nitrogen mineralization by degrading urea, and phosphatases on the other hand (Acidic or alkaline) are indicators of overall agricultural and forest management and their pollution levels.

The currents methods available for estimation of available phosphorus are not truly reflecting the soil profile as expected by the scientific community. The major hurdle of phosphatase analysis is the ambiguity of the source of the enzyme (free or intracellular). Free enzymes also contribute toward soil health but require a stabilizing environment in soil. The main hindrance is a method that will aid in direct visualization or assay of free enzyme in soil that is the most vital challenge to be met. The most important need of soil biology is the development of methods that would assess and distinguish important soil enzymes viz. lipases, dehydrogenases, catalases, and ureases in whether extracellular or intracellular.

References

Atalla, S.M.M., El Gamal, N.G., 2020. Production and characterization of xylanase from pomegranate peel by *Chaetomium globosum* and its application on bean under greenhouse condition. Bull. Natl. Res. Cent. 44, 104. https://doi.org/10.1186/s42269-020-00361-5.

Babu, S.V., Triveni, S., Reddy, R.S., Sathyanarayana, J., 2017. Int.Influence of application of different formulations of phosphate solubilizing biofertilizers on soil enzymes in maize crop. J. Curr. Microbiol. App. Sci. 6 (12), 377.

Burak, K., 2020. Importance of urease activity in soil. In: Conference: V. International Scientific and Vocational Studies Congress – Science and Health (BILMES SH 2020).

Dakshayani, S.D.S., Sharathchandra, R.G., Jogaiah, S., 2015. Screening of endophytic fungi for their ability to produce extracellular cellulases. Int. J. Pharm. Pharmaceut. Sci. 7, 205–211.

Deng, S., Tabatabai, M., 1997. Effect of tillage and residue management on enzyme activities in soils: III. Phosphatases and arylsulfatase. Biol. Fertil. Soils 24, 141–146. https://doi.org/10.1007/s003740050222.

Eivazi, F., Tabatabai, M.A., 1977. Phosphatases in soils. Soil Biol. Biochem. 9 (3), 167–172.

El-Sheikh, E.A., Kamita, S.G., Vu, K., Hammock, B.D., 2011. Improved insecticidal efficacy of a recombinant baculovirus expressing mutated JH esterase from *Manduca sexta*. Biol. Contr. 58, 354–361.

Ferraz de Almeida, R., Naves, E.R., Pinheiro da Mota, R., 2015. Soil quality: enzymatic activity of soil β-glucosidase. Global J. Agricul. Res. Rev. 3 (2), 146–150.

Hanzlik, T.N., Hammock, B.D., 1988. Characterization of juvenile hormone hydrolysis in early larval development of Trichoplusia ni. Arch. Insect Biochem. Physiol. 9, 135–156.

Jarvan, M., Edesi, L., Adamson, A., Vosa, T., 2014. Soil microbial communities and dehydrogenase activity depending on farming systems. Environ. Plant Soil 60, 459–463.

Jogaiah, S., Abdelrahman, M., Tran, L.-S.P., Ito, S.-I., 2018. Different mechanisms of *Trichoderma virens*-mediated resistance in tomato against Fusarium wilt involve the jasmonic and salicylic acid pathways. Mol. Plant Pathol. 19, 870–882.

Konappa, N., Udayashankar, A.C., Dhamodaran, N., Krishnamurthy, S., Jagannath, S., Uzma, F., Pradeep, C.K., De Britto, S., Chowdappa, S., Jogaiah, S., 2021. Ameliorated antibacterial and antioxidant properties by *Trichoderma harzianum* mediated green synthesis of silver nanoparticles. Biomolecules 11, 535.

Kramer, K.J., Muthukrishnan, S., 1997. Insect chitinases: molecular biology and potential use as biopesticides. Insect Biochem. Mol. Biol. 27 (11), 887–900. https://doi.org/10.1016/s0965-1748(97)00078-7. PMID: 9501415.

Kumar, S., Chaudhuri, S., Maiti, S.K., 2013. Soil dehydrogenase enzyme activity in natural and mine soil—a review. Environ Ecol. Res. 30, 898–906.

Małachowska-Jutsz, A., Matyja, K., 2019. Discussion on methods of soil dehydrogenase determination. Int. J. Environ. Sci. Technol. 16, 7777–7790. https://doi.org/10.1007/s13762-019-02375-7.

Margalef, O., Sardans, J., Fernández-Martínez, M., Molowny-Horas, M., Janssens, I.A., Ciasis, P., Goll, D., Richter, A., Obersteiner, M., Asensio, D., Penuelas, J., 2017. Global patterns of phosphatase activity in natural soils. Sci. Rep. 7, 1337. https://doi.org/10.1038/s41598-017-01418-8.

Maseko, S.T., Dakora, F.D., 2013. Rhizosphere acid and alkaline phosphatase activity as a marker of P nutrition in nodulated Cyclopia and Aspalathus species in the Cape fynbos of South Africa. South Afr. J. Bot. 89, 289–295.

Nalini, P., Ellaiah, P., Prabhakar, T., Girijasankar, G., 2015. Microbial alkaline phosphatases in bioprocessing. Int. J. Curr. Microbiol. App. Sci. 4 (3), 384–396.

Nannipieri, P., Giagnoni, L., Landi, L., Renella, G.E.K., Cunemann, B., 2015. Chapter 9 role of phosphatase enzymes in soil. In: Phosphorus in Action. Soil Biology, vol. 26, pp. 2015–2243. https://doi.org/10.1007/978-3-642-15271-9_9. Springer-Verlag Berlin Heidelberg 2011.

Piotrowska-Długosz, A., 2019. Significance of enzymes and their application in agriculture. In: Husain, Q., Ullah, M. (Eds.), Biocatalysis. Springer, Cham. https://doi.org/10.1007/978-3-030-25023-2.

Pushpalatha, H.G., Jogaiah, S., Ashok, S., Geetha, N.P., Kini, K.R., Shekar Shetty, H., 2013. Association between accumulation of allene oxide synthase activity and development of resistance against downy mildew disease of pearl millet. Mol. Biol. Rep. 40, 6821–6829.

Satapute, P., Milan, V.K., Shivakantkumar, S.A., Jogaiah, S., 2019. Influence of triazole pesticides on tillage soil microbial populations and metabolic changes. Sci. Total Environ. 651, 2334–2344.

Sharma, S.B., Sayyed, R.Z., Trivedi, M.H., Gobi, T.A., 2013. Phosphate solubilizing microbes: sustainable approach for managing phosphorus deficiency in agricultural soils. SpringerPlus 2, 587.

Singh, M., Kumar, A., Singh, R., Pandey, K.D., 2017. Endophytic bacteria: a new source of bioactive compounds. 3 Biotech 7 (5), 315. https://doi.org/10.1007/s13205-017-0942-z.

Sudisha, S., Niranjana, S.R., Umesha, S., Prakash, H.S., Shekar Shetty, H., 2006. Transmission of seed-borne infection of muskmelon by *Didymella bryoniae* and effect of seed treatments on disease incidence and fruit yield. Biol. Contr. 37, 196–205.

Tan, A., Tanaka, H., Tamura, T., Shiotsuki, T., 2005. Precocious metamorphosis in transgenic silkworms overexpressing juvenile hormone esterase. In: Proceedings of the National Academy of Sciences of the United States of America, vol. 102, pp. 11751–11756.

Wang, J.B., Chen, Z.H., Chen, I.J., Zhu, A.N., Wu, Z.J., 2011. Surface soil Phosphorus and surface activities affected by Tillage and crop residue input amounts. Plant Soil Environ. 57 (6), 251–257.

Zeng, X., Zhou, Y., Zhu, Z., Zhu, H., Wang, S., Di, H., Wang, Z., 2016. Effect on soil properties of BcWRKY1 transgenic maize with enhanced salinity tolerance. Int. J. Genom. 6019046. https://doi.org/10.1155/2016/6019046.

SUBCHAPTER

Production of enzyme biofertilizer—phosphatase 11.1

Before you begin

Timing: 13 days.

Media preparation

1. **Preparation of Czapeckdox agar plates**
   Dissolve sucrose 30.0 g, $NaNO_3$ 3.0 g, $MgSO_4$ 0.5 g, KCl 0.5 g, $FeSO_4$ 0.01 g, KH_2PO_4 1.0 g, and distilled water 1000 mL and adjust the pH to 5.50. Add 20 g agar agar powder and homogenize.
   Sterilize the medium and prepare Czapeckdox agar plates for inoculation.
2. **Screening medium for phosphatase (g/L)**
   Prepare 300 mL of nutrient broth and add 300 mL of potato dextrose broth. Add 0.1 g of P-nitrophenylphosphate. This screening medium is then added to a 500 mL flask and sterilized.
3. **Preparation of Buffer**
   The modified universal buffer is prepared by adding 12.1 g of tris-hydrochloric aminomethane, 11.6 g of malic acid, 14. 0 g of citric acid, 6.3 g of boric acid in water, and 488 mL of 0.1 M NaOH (pH 5.0).
4. ***p*-nitrophenylphosphate solution**
   115 mM *p*-nitrophenylphosphate is prepared in 50 mL.
5. **0.5 M NaOH Solution**
   Dissolve 2 g of NaOH in 100 mL of distilled water.

Key resources table

Note that not all areas will be used in every protocol.

Reagent or resource	Source	Identifier
Biological samples		
Fungal culture		
Aspergillus ficuum (Han and Gallagher, 1987) Can isolate microbes from rhizosphere or any novel source (Olayide et al., 2018)	Isolated or procured	

Materials and equipment

- **Materials:** Erlenmeyer flasks, volumetric flasks, conical flasks, funnel, test tubes, test tube stand, Pipettes, Whatman filter paper 1, and pipette stand.
- **Equipment:** Weighing balance, autoclave, shaking incubator, shaking water bath, boiling water bath, centrifuge and visible spectrophotometer.
- **Alternatives**: Colorimeter can be used instead of spectrophotometers.

Step-by-step method details

Preparation of inoculum

Timing: Day 1–Day 7

1. Take the pure culture of the selected fungi.
2. Inoculate on the Czapeckdox plates.
3. After incubation for 7 days scrap the spores and suspend them in a minimum amount of spore suspension medium.
4. After all the spores are transferred, dilute the suspension in an adequate manner to receive 10^8 spores/mL.

Note: (If Bacteria is used for production then Day 1–Day 2)

Preparation of screening medium

Timing: Day 7

1. The screening medium is prepared and sterilized.

Inoculation of the screening medium

Timing: Day 8

1. The autoclaved screening medium is allowed to cool and inoculated with 0.5 mL of the above spore suspension as inoculum under aseptic conditions. The medium is incubated at 27°C for 5 days.
2. The unbaffled Erlenmeyer flask serves as a small-scale fermentor.
3. Place in shaking incubator at 150 rpm at 27°C for 5 days.
4. Monitoring should be done by withdrawing samples after a fixed time interval (e.g., 12 h).
5. Centrifuge at 1500 rpm for 20 min.
6. The supernatant is used as a crude enzyme source for assay.

Pause Point: All the incubation periods where the organism is given optimum conditions for growth and enzyme productions are the pause points.

Termination of fermentation

Timing: Day 13

1. Fermentation is stopped, and final enzyme production is determined by assay.
2. Enzyme assay Tabatabai and Brimner (1969).
3. Add 0.1 mL of crude enzyme.
4. Add 1 mL of modified universal buffer and 0.2 mL *p*-nitrophenylphosphate.
5. Incubate the mixture for 1 h at 37°C.
6. Add 0.1 mL of 0.5 M NaOH to terminate the reaction.
7. Centrifuge for 15 min at 10,000 rpm immediately.
8. Read the absorbance at 405 nm for determining p-nitrophenol released.
9. Prepare a standard curve of p-nitrophenol for calculating the quantity of enzyme.

Note-One unit of enzyme activity was defined as the amount of enzyme required to liberate 1 mM of *p*-nitrophenol per mL per min under assay conditions.

Expected outcomes

Depending on the potential of the microbe selected for enzyme production, the enzyme is produced on small scale.

Supplement the crude enzyme in the soil and observe the growth characteristics of control plant and plant grown in enzyme biofertilizer remediated soil.

Quantification and statistical analysis

The standard protocol for assay will be followed.

Advantages

The procedure is too simple.

No expertise is required.

Limitations

Success of the experiment depends on the potential of the fermenting microorganism chosen.

In the case of molds, the time required for enzyme production is prolonged due to the long incubation period.

Optimization and troubleshooting

Contamination is the most common problem encountered in fermentation that affects enzyme production.

Enzyme activity is not detected due to inhibitors produced by contaminants. Medium precipitates after sterilization, and no growth occurs. Medium is proper but no growth occurs. All these reasons hamper enzyme production, and no activity is registered in the crude sample.

Potential solution to optimize the procedure

If contamination occurs, there is no alternative then to discard and start a fresh.

Strict sterilization and aseptic transfers are the measures that can avoid contaminations.

Di-potassium hydrogen phosphate has to be added as a solution after all the ingredients are dissolved in a partial volume of water.

The medium should be properly cooled before inoculation.

Many times, the enzymes are sensitive to Fe salts. If all parameters are proper, then Fe salt can be avoided to cross-check to study its inhibitory effect on the enzyme.

Safety considerations and standards

After the assay is complete, the fermentation remnants should be discarded after proper decontamination.

Alternative methods/procedures

Standard protocol for phosphatase assay is used.

References

Han, Y.W., Gallagher, D.J., 1987. Phosphatase production by *Aspergillusficuum*. J. Industr. Microbiol. 1, 295–301. https://doi.org/10.1007/BF01569307.

Olayide, F.O., Olushina, O.A., Miriam, N.E., Foluke, O.O., 2018. Production of phosphatase by microorganisms isolated from discolored painted walls in a typical tropical environment: a Non-Parametric analysis. Arab J. Bas. Appl. Sci. 25 (3), 111–121. https://doi.org/10.1080/25765299.2018.1527277.

Tabatabai, M., Bremmer, J., 1969. Use of p-nitrophenylphosphate for assay of soil phosphatase activity. Soil Biol. Biochem. 1, 301–307.

CHAPTER

Biotechnological applications of enzymes and future prospective

12

12.1 Introduction

All enzymes within themselves are magical molecules bestowed with magnificent biotransformation capability. They have been recognized for their catalytic virtue since industrialization crept into human civilization. With increasing modernization in human lifestyles, the industrial dependency is proportionally increasing upon enzymes. The actions of enzymes are nowadays gaining popularity and industrial acceptance due to their mild action-specific results, eco-friendly nature, and harmless effectiveness upon the consumers. The chemical-based products are now facing consumer rejection even though they compete in the market on grounds of low cost. Consumers are analyzing enzyme-based products or processes that influence the market demand directly. Enzymes are involved in almost all walks of life as shown in Fig. 12.1.

12.2 Enzymes in medical field

Although enzymes have gathered attention in the global market on account of their technological applications in industrial processes by paper, leather, etc. yet their role in the medical field is also worth mentioning.

Medical employability of enzyme is still in the Juvenile stage as it suffers from a major setback like the requirement of ultrapure enzymes.

In the medical field, enzymes have two roles to perform:

1. In the detection of diseases
2. As a therapeutic agent

12.2.1 Enzymes and diagnosis

In the medical field, diagnosis of the disease is critically important. There are various parameters in which various metabolites are traced. The enzymes used in the diagnostics are either directly used or is one of the components of the detecting system.

Protocols and Applications in Enzymology. https://doi.org/10.1016/B978-0-323-91268-6.00006-5

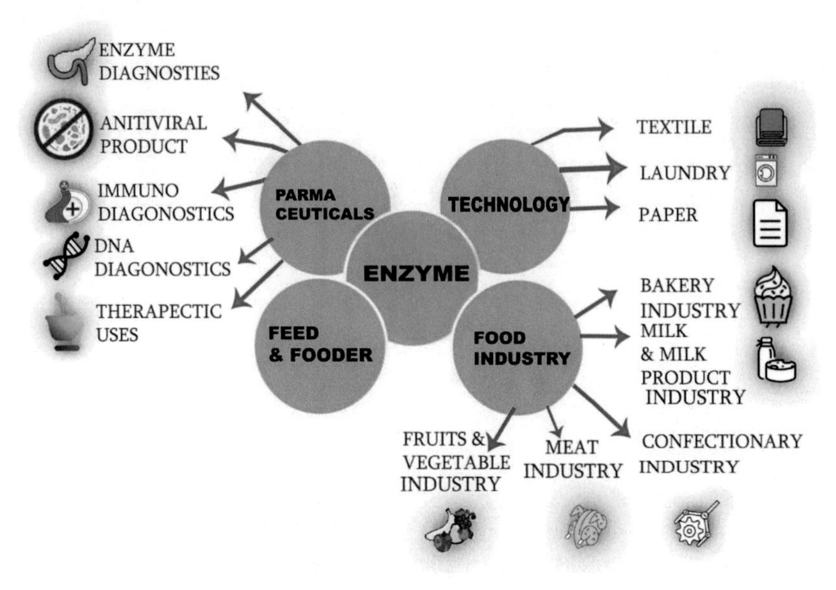

FIGURE 12.1

Intervention of enzymes in different walks of human life.

The principle used in enzyme-based diagnosis is to detect a key metabolite indicator of the disease or disorder. Specific enzymes are used to transform this metabolite into a product that is measurable or visually detectable (Fig. 12.2). Such a detection process simplifies the diagnostic procedure as it overcomes the necessity to purify or separate target diagnostic metabolite from the patient's sample.

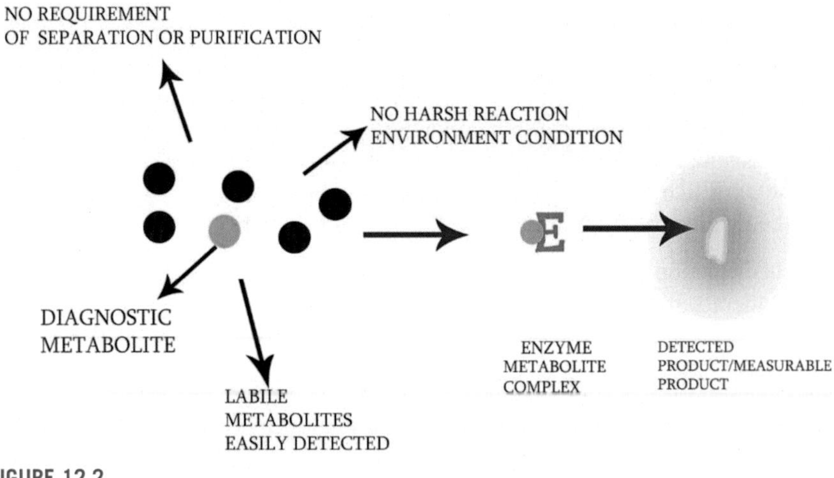

FIGURE 12.2

The principle of enzyme-based diagnostics.

This method also remains unaffected by the presence of other metabolites due to the virtue specificity of the enzyme for the target metabolite. Enzymes as detecting agents confer the ability to detect labile metabolites without generating any harsh detecting/reaction environment as is the case of chemical detection agents.

12.2.2 Desirable features of diagnostic enzyme

1. It should have equilibrium toward the right side so that a large amount of product will be formed and will make the detection easier.
2. If such a condition is not prevailing, a secondary reaction has to be supplemented with a second enzyme where a small amount of product can be retransformed to a secondary product in a respectable concentration or a metabolite that can be easily detected.
3. Enzymes selected for diagnostic should be stable, resistant toward the environment, and accompanying metabolites that would hamper its activity otherwise.
4. Extremely purified enzymes are required in diagnostic protocols, traces of contaminants can ruin the detection strategy due to virtue of enzyme sensitivity.
5. Absolute specificity is desirable as it should not convert accompanying metabolites to similar products.

The compounds that are generally detected are acetic acid, alcohol, heavy metals, proteins, and even enzymes. There are three intense fields in diagnostics where enzymes remarkably contribute.

12.2.3 In disease diagnosis

As explained in the introduction, a disease indicative metabolite can be traced if it happens to be a substrate for a diagnostic enzyme. The product detection can be coupled with available colorimetric, spectrophotometric, or fluorescence-based detection methods. The examples of such basic enzymes used as marker enzymes since the past are given in Table 12.1.

In many instances, the accompanying product is much more accessible for easy detection that renders the process feasible for undetectable products. The metabolic role of glutamate dehydrogenase is elaborated in Fig. 12.3. This enzyme is used for the detection of liver damage.

Many enzymes in blood prove as a wonderful indicator system for the normal functioning of vital organs since the past. Liver necrosis and alcoholism have been investigated in relation to one another, and biopsy revealed the levels of glutamate dehydrogenase, serum aspartate and alanine transferase (SGOT and SGPT), ornithine carbamoyltransferase, and γ-glutamyltranspeptidase proportionally increased with the progression of necrosis. All other enzymes exhibited overlapped values between actual patients and controls. Only glutamate dehydrogenase levels clearly demarcated the level of necrosis and rise in concentration. As the difference registered was two and half times more, it was used as a successful maker (Van Waes and Lieber, 1977).

Table 12.1 Marker enzymes in disease diagnosis.

Disease	Marker enzyme	Ref.
Bone and hepatobiliary diseases	Alkaline phosphatase	Lone et al. (2003)
Rheumatoid arthritis	Alkaline phosphatase	Chandrakar et al. (2017)
Bone metastasis		Jin et al. (2015)
Myocardial infarction and muscle diseases	Creatine kinase	Aksenova et al. (2000)
Acute infection	Serum cholinesterase	Ellis (2005)
Muscular dystrophy Chronic renal disease Pregnancy Insecticide intoxication		
Pancreatitis	Amylase	Gupta et al. (2001)
Hepatobiliary disease Alcoholism	Glutamyltransferase	Kaneko (1989)
Breast cancer	Cathepsin D	Schlimpert et al. (2020)

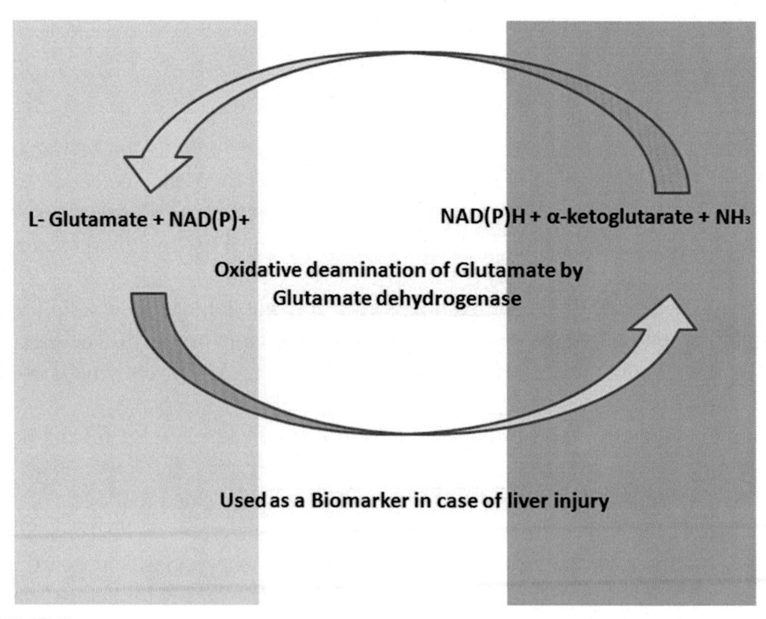

FIGURE 12.3

The function of glutamate dehydrogenase.

12.2.4 Coupled enzyme assay

The coupled enzyme assay is best exemplified by a reaction converting phosphoenolpyruvate to pyruvate. This reaction is catalyzed by enzyme 1 pyruvate kinase. Clinically, the rate of the reaction catalyzed by enzyme 1 is to be assayed for which a measurable and detectable metabolite is required. The subsequent reaction is the conversion of pyruvate to lactate that is associated with oxidation of NADH that is easily measurable by UV spectrophotometry (Kiianitsa et al., 2003). Fig. 12.4 explains the detection of activity of phosphoenol pyruvate.

There are various challenges in successfully accomplishing the coupled assays, as two reactions are simultaneously addressed one after the other in the same reaction medium. The general issues encountered are differences in the optimal conditions of both enzymes like pH change. Studies are now focused to handle and improvise these coupling outputs (Moisa et al., 2020).

12.2.5 Immunological reactions

This remains an unparalleled diagnostic field for any other diagnostic tool rather than enzymes for intervention. The coupling of diagnostic enzymes and immunological reactions are referred as enzyme immunoassays. The basic components of immunoassay are shown in Fig. 12.5.

The types of enzyme immunoassay are as follows.

Homogeneous immunoassay: There is no need to separate Ag—Ab complex before analysis.

Easier and faster to perform. Used for detection of small analytes, drugs, etc.

Heterogeneous assays: Bound and free antibody must be separated before measuring the label.

Fig. 12.6A and B explains the competitive and noncompetitive immunoassay.

12.2.6 DNA-based diagnostic

The entire identity test is based on DNA homology wherever a target DNA is to be amplified. The protocol calls for enzyme intervention. The PCR-based methods and

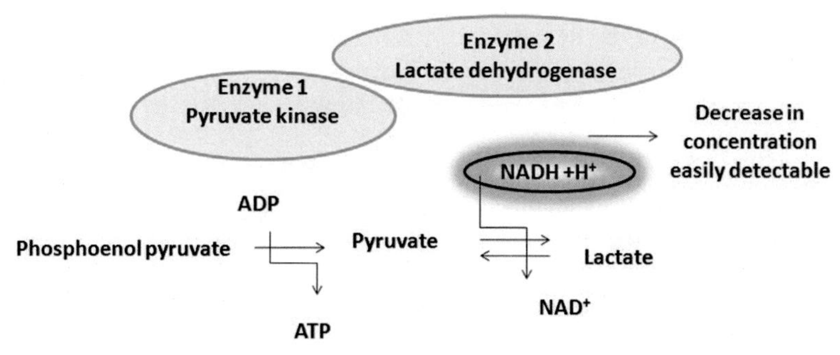

FIGURE 12.4

Coupled enzyme assay.

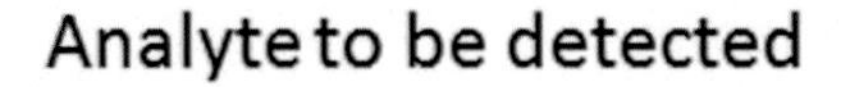

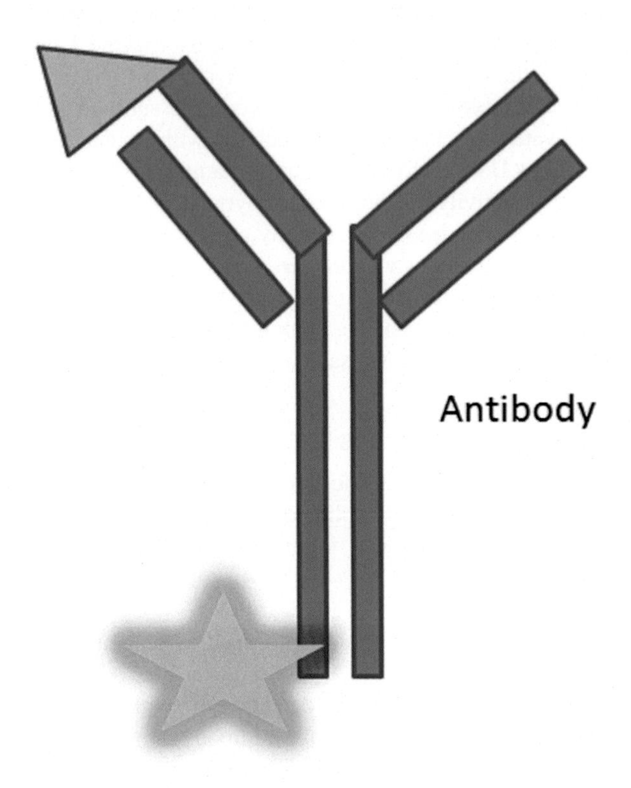

FIGURE 12.5

Components of enzyme immunoassay.

established methods in the department of forensics. Following diseases are diagnosed by different enzymes as enlisted in Table 12.2.

12.2.7 Therapeutic enzymes

Direct usage of enzymes as therapeutic agents is a field with little exposure. Well known for their catalytic power, enzymes offer hindrance due to their macromolecular nature in direct ingestion and dissemination in various body parts.

Movement of the enzyme toward the target site within the body.

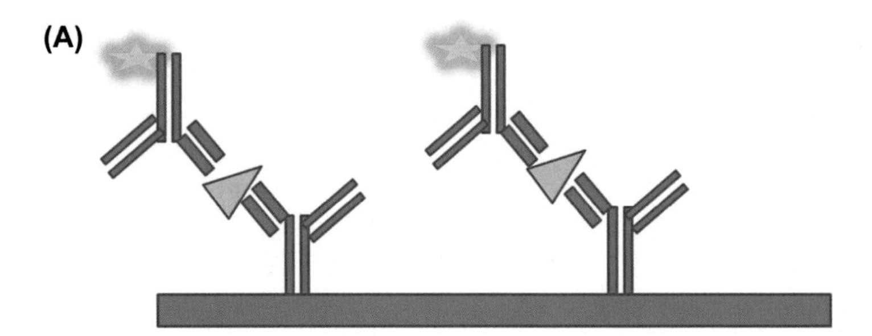

Two sites non competitive immuno assay: The ELISA technique uses a bound antibody, a labelled antibody and the metabolite is sandwitched between the two antibodies. The signal attached to the detector antibody reveals the positivity of the test.

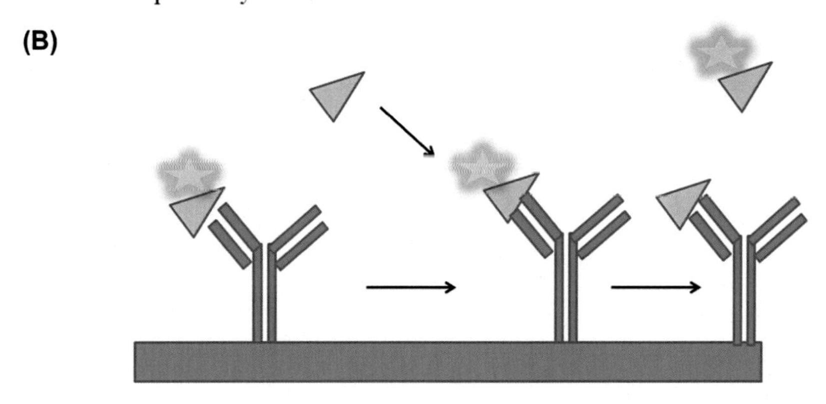

Competitive homogeneous immuno assay: The unlabelled analyte competes with the labelled bound analyte and replaces it. The displaced labelled analyte is then detected.

FIGURE 12.6

(A) Competitive immunoassay and (B) Noncompetitive immunoassay.

Enzymes can be recognized as foreign molecules and can trigger immunogenic responses.

Circulatory system does not allow a long-term circulation of foreign molecules.

On the contrary, enzymes have been successfully implemented as topical therapeutic agents, anticancer agents, in genetic disorders and deficiency diseases. Following enzymes are very popularly used as therapeutic agents. Table 12.3 enlists the enzymes used as therapeutic agents themselves.

Table 12.2 Biosensor enzyme-based diagnostic methods.

Enzymes	Application in detection	Metabolite detected
Lactate oxidase	Ischemic myocardium	Lactate (Rathee et al., 2016)
Glucose oxidase	Diabetes mellitus	Glucose (Ridhuan et al., 2018)
Glutamate oxidase	Neuropathology	Glutamate (Soldatkina et al., 2017)
Oxalate oxidase	Kidney stone	Oxalate (Hu et al., 2015)
Creatinine amidohydrolase, Creatine amidinohydrolase, Sarcosine oxidase	Kidney disorder	Creatinine and creatine (Kim et al., 2013) Pundir et al. (2018)
Urate oxidase	Kidney disorder	Uric acid (Falohun and McShane, 2020)
Cholesterol oxidase	Cholesterol level	Cholesterol (Lin et al., 2016)
Creatininase	Creatine Creatinine levels in urine	Creatinine (Duong and Rhee, 2017)
Lactate dehydrogenase	Hemorrhage Respiratory failure Hepatic disease Sepsis Tissue hypoxia	Lactate (Rathee et al., 2016)
AST ALT	Liver diseases and myocardial infarction	Aspartate transaminase Alanine transaminase (Song et al., 2009)
Alkaline phosphatase	Bone diseases	Marker enzyme (Sappia et al., 2019)
Polyaniline (PANI) nanowire biosensors	Cardiovascular disease	Creatinine kinase myoglobin (Myo) cardiac troponin (Lee et al., 2012)
Acid phosphatase Glucose oxidase	Pesticide detection	Organophosphorus Carbamic acid (Mazzei et al., 1996)

12.3 The biological detergents: a revolution in laundry industry

Enzymes find a wide variety of applications in the market. The laundry industry has the second-highest demand after the food industry for enzyme employability (Takimura et al., 2007). The global market will evidence a rise of \$7.0 billion by 2023 from \$5.5 billion in 2018 with a compound annual growth rate (CAGR) of 4.9% for the period 2018–23 according to the Global Markets for Enzymes in Industrial Applications. The analysis of enzyme demand reveals that of the total enzyme, and

Table 12.3 Therapeutic applications of some important enzymes.

Enzymes	Therapeutic applications
Bromelain, Renin, and Papain	To cure digestive disorder Reduces pain in osteoarthritis Antiinflammatory agent Removes dead tissues
Catalase	Antioxidant treatment
Cellulase	Digestive supplement
Ribonuclease	Antiviral agent
Trypsin	Antiinflammatory
Collagenase	Skin ulcers
Lipase	Muscular spasm
Ficin and Papain	Deworming intestine
Asparaginase Glutaminase	Tumor treatment
Streptokinase Urokinase	Treats blood clots generated in the circulatory system

protease enjoys the major market demand (60%–65%). Protease offers a wide range of employability due to its activity in diverse pH and temperature conditions (Gupta et al., 2002; Rao et al., 1998; Pushpalatha et al., 2013). The benefits of enzyme-based detergents are shown in Fig. 12.7.

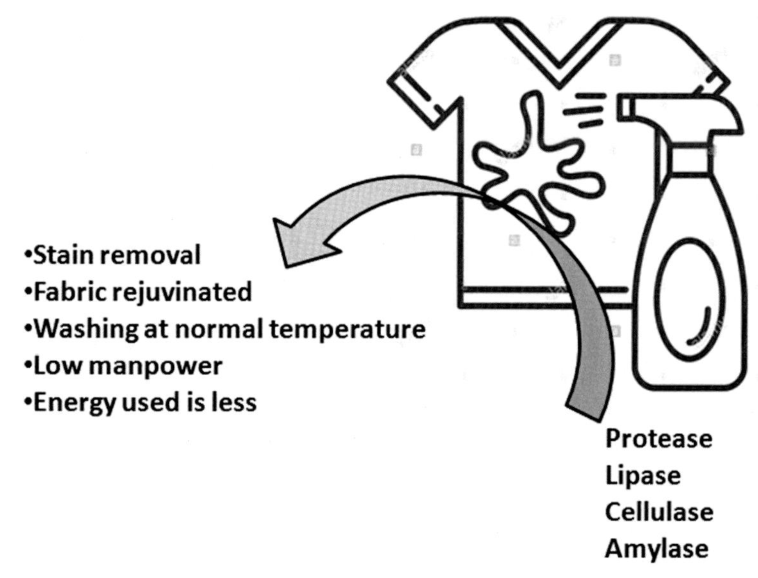

FIGURE 12.7

Advantages of enzyme-based detergents.

Conventional detergents are the most rejected products nowadays. They suffer major drawbacks like

- Long hours of soaking
- Weakens fabric and stitching
- Excess water is required for cleaning
- Higher degree of agitation and more manpower is required for removal of chemical detergent

The chemical nature damages the skin of the user and above all after use terminates in environmental pollution. Biological detergents mitigate all the above issues and confer added advantages of eco-friendly and user-friendly benefits.

The history of biological detergents reveals the use of crude enzymes that accompany undesirable impurities causing allergy to the user. A modification is introduced where it is encapsulated in granular form avoiding the dust effects. Enzymes are provided a protective shield of carboxymethyl cellulose. The enzyme is trapped with preservative, polyethylene glycol, and hydrophilic material with a hydrophobic coat. The enzymes popularly used in the laundry industry are amylases, lipases, proteases, and cellulases.

12.3.1 Proteases

Proteases help to remove protein stains that adhere intimately to the fabric fibers. It successfully removes resistant stains without damaging the fiber in contrast to the chemical detergents. The commercially used proteases provide a wide range of pH and temperature to be used in conditions. The enzymes exhibit cleansing and destaining activity from 10°C to 65°C and pH 12. Protease in combination with surfactants provides much more satisfactory removal of protein stains.

Protease show activity within various pH ranges, and their classification can be done as acidophilic, neutrophilic, and alkaliphilic enzymes (Garcia-Carreno, 1991). According to IUB, there are categories of proteases:

1. Serine I
2. Serine II
3. Cysteine (plants)
4. Cysteine (animals)
5. Aspartic
6. Metalloproteases

On the basis of internal peptide bond cleavage and active site, nature proteases are classified as follows:

1. Serine proteases (EC.3.4.21)
2. Cysteine proteases (EC.3.4.22)

3. Aspartic proteases (EC.3.4.23)
4. Metalloproteases (EC.3.4.24)

The serine proteases are the most eligible enzymes for the laundry industry. The cysteine proteases face the drawback of oxidation in the presence of bleaching agents. The aspartic proteases are active at acidic pH and are more useful in baking and food industry. The metalloproteases lose their activity in the detergent form due to the loss of metal ions due to hydroxyl ion or water softening agents.

12.3.2 Lipase

This is one of the important enzymes used in the laundry industry. It helps to remove the lipid stains. The most attractive asset lipase offers is the removal of greasy stains at a normal temperature that is the advantage of enzyme-based detergents.

Selection of lipase for laundry applications demands qualities like relative specificity towards substrate, alkaliphilic nature, thermostablility, thermophilic nature, stability in the presence of other components of detergent like protease and surfactants.

Chemical detergents fall short in removing stains at normal temperatures. The washing practice at elevated temperatures causes harm to fabric apart from making it weak. Usage of lipase bridges this gap by removal of lipid stain at normal temperature. Stable alkaliphilic lipases are the most desirable enzymes in the market (Hasan et al., 2006).

12.3.3 Amylases

Amylases are commonly used in dishwashing detergents for the removal of starchy stick-outs on utensils. Proteases are also used but enzyme compatibility is checked before their combined usage. Following protease, nearly 25% of the market is for the amylase enzyme. It is also used in laundry detergents to remove carbohydrate stains. The challenge is to procure a protease tolerant enzyme that can also withstand the other components in the detergent formulation.

12.3.4 Cellulases

Cellulases help to modify the fiber texture of the clothes washed. Repeated washings develop microfibril branching from the main fabric and make the cloth look dull and overused. Cellulase removes these microfibrils and improves the overall look of the fabric and maintains the health of the cloth fabric. A comparative account of major ingredients of enzyme-based and chemical detergents is enlisted in Table 12.4.

Table 12.4 Enzyme-based detergents and commercial detergents.

Enzyme based detergents	Role in washing	Commercial detergent	Role in washing
Anionic surfactants	To remove dirt	Sodium alkane sulfonate	Surfactant to remove dirt
Nonionic surfactants	To reduce soap scum	Sodium tripolyphosphates	Water softener
Soap	Cleansing	Sodium perborate tetrahydrate	Oxidizing agents
Sequestrants	Reduces ions to enhance cleaning	Sodium alkane carboxylate	Soap
Enzyme	Specific target toward stains	Carboxy methyl cellulose	Dirt suspending agents
Activator	Provides effective cleaning at low temperatures	Sodium metasilicate	Binder and dirt loosener
Perfume	Imparts fragrance	Fluorescent brighteners	Whiteners
Brighteners	Dyes absorbed by the fabric reflects light at the bluish side		
Sodium sulfate	Acts as filler and provides powdery body to detergent		

12.4 Conclusion

The most promising horizons of microbial technology lie in the prospects of enzyme applications. The innovations in industrial processes have elevated the enzyme production level without compromise in the economy. The medical field is on the verge of patenting maximum enzyme-based diagnostics. Initially, kits were used but nowadays due to lack of accuracy and precision in the detection of traces of metabolites, enzymes were the most focused targets. Recently, biosensor-based direct detection of metabolites is much in trend rather than reaction-based diagnostics.

Biosensors offer a great deal of flexibility, economy, reusability, and short-timed analysis. Enzymes have to be made more robust and stable as per the environmental conditions of the activity and should be designed according to protocol. The enzymes also extend a new era of therapy based on the direct use of enzymes as therapeutic agents. The diseases offering challenges of being incurable are now being approached through enzyme therapy. Genetic disorders also are the targets to be resolved by enzyme therapy. Technological applications mainly desire to focus on the development of less sensitive and more specific tailor-made enzymes. The wastes are used as substrates to synthesize enzymes using green technology. In the laundry industry, all the modern screening techniques are employed to screen a novel enzyme that desired thermal stability and tolerance toward alkaline pH.

Enzymes with these features are produced but the cost of production makes them available at a higher price. The need of the hour is to design industrial processes utilizing low-cost substrates and technologies to create a sustainable and eco-friendly tool.

References

Aksenova, M., Butterfield, D.A., Markesbery, W.R., 2000. Oxidative modification of creatine kinase BB in Alzheimer's disease brain. J. Neurochem. 74, 2520–2527.

Chandrakar, B.L., Sharma, H.C., Chandrakar, K.C., 2017. Activity of serum alkaline phosphatase in rheumatoid arthritis for diagnosis and management. Int. J. Med. Res. Prof. 3 (3), 281–284.

Duong, H.D., Rhee, J.I., 2017. Development of ratiometric fluorescent biosensors for the determination of creatine and creatinine in urine. Sensors 17 (11), 2570. https://doi.org/10.3390/s17112570.

Ellis, J.M., 2005. Cholinesterase inhibitors in the treatment of dementia. J. Am. Osteopath. Assoc. 105, 145–158.

Falohun, T., McShane, M.J., 2020. An optical urate biosensor based on urate oxidase and long-lifetime metalloporphyrins. Sensors 20 (4), 959. https://doi.org/10.3390/s20040959.

Garcia-Carreno, F.L., 1991. Protease inhibition in theory and practice. Bitech. Edu. 3, 145–150.

Gupta, K.B., Ghalaut, V., Gupta, R., Tandon, S., Prakash, P., 2001. Estimation of serum and pleural fluid amylase and iso-enzyme in cases of malignant pleural effusion. Indian J. Tubercul. 48, 87.

Gupta, R., Beg, Q.K., Lorenz, P., 2002. Bacterial alkaline proteases: molecular approaches and industrial application. Appl. Microbiol. Biotechnol. 59, 15–32.

Hasan, F., Shah, A.A., Hameed, A., 2006. Industrial applications of microbial lipases. Enzym. Microb. Technol. 39, 235–251.

Hu, Y., Xiang, M., Jin, C., Chen, Y., 2015. Characteristics and heterologous expressions of oxalate degrading enzymes "oxalate oxidases" and their applications on immobilization, oxalate detection, and medical usage potential. J. Biotech Res. 6, 63–75.

Jin, Y., Yuan, M.Q., Chen, J.Q., Zhang, Y.P., 2015. Serum alkaline phosphatase predicts survival outcomes in patients with skeletal metastatic nasopharyngeal carcinoma. Clinics 70 (4), 264–272. https://doi.org/10.6061/clinics/2015(04)08.

Kaneko, J.J., 1989. Clinical Biochemistry of Domestic Animals, fourth ed. Academic Press, New York, California, USA., p. 898

Kiianitsa, K., Solinger, J.A., Heyer, W.-D., 2003. NADH-coupled microplate photometric assay for kinetic studies of ATP-hydrolyzing enzymes with low and high specific activities. Anal. Biochem. 321 (2), 266–271.

Kim, H.C., Oh, S.M., Pan, W.H., Ueshima, H., Gu, D., Chuang, S.Y., Fujiyoshi, A., Li, Y., Zhao, L., Suh, I., 2013. East asian network for stroke prevention study group. Association between alanine aminotransferase and intracerebral hemorrhage in east asian populations. Neuroepidemiology 41 (2), 131–138. https://doi.org/10.1159/000353186.

Lee, I., Xiliang, L., Jiyong, H., Cui, H., Yun, X.M., 2012. Detection of cardiac biomarkers using single polyaniline nanowire-based conductometric biosensors. Biosensors 2, 205–220. https://doi.org/10.3390/bios2020205.

Lin, X., Ni, Y., Kokot, S., 2016. Electrochemical cholesterol sensor based on cholesterol oxidase and MoS2-AuNPs modified glassy carbon electrode. Sensor. Actuator. B Chem. 233, 100–106.

Lone, M.A., Wahid, A., Saleem, S.M., Koul, P., Nabi Dhobi, G.H., Shahnawaz, A., 2003. Alkaline phosphatase in pleural effusions. Indian J. Chest Dis. Allied Sci. 45, 161–163.

Mazzei, F., Botrè, F., Botrè, C., 1996. Acid phosphatase/glucose oxidase-based biosensors for the determination of pesticides. Anal. Chim. Acta 336, 67–75.

Moisa, M.E., Amariei, D.A., Nagy, E.Z.A., Szarvas, N., Tosa, M.I., Paizs, C., Bencze, L.C., 2020. Fluorescent enzyme-coupled activity assay for phenylalanine ammonia-lyases. Sci. Rep. 10, 18418.

Pundir, C.S., Kumar, P., Ranjana, J., 2018. Biosensing methods for determination of creatinine: a review. Biosens. Bioelectron. 126. https://doi.org/10.1016/j.bios.2018.11.031.

Pushpalatha, H.G., Jogaiah, S., Ashok, S., Geetha, N.P., Kini, K.R., Shekar Shetty, H., 2013. Association between accumulation of allene oxide synthase activity and development of resistance against downy mildew disease of pearl millet. Mol. Biol. Rep. 40, 6821–6829.

Rao, M.B., Tanksale, A.M., Ghate, M.S., Deshpande, V.V., 1998. Molecular and Biotechnological aspects of microbial proteases. Microbiol. Mol. Biol. Rev. 62, 597–635.

Rathee, K., Dhull, V., Dhull, R., Singh, S., 2016. Biosensors based on electrochemical lactate detection: a comprehensive review. Biochem. Biophys. Rep. 5, 35–54.

Ridhuan, N.S., Abdul Razak, K., Lockman, Z., 2018. Fabrication and characterization of glucose biosensors by using hydrothermally grown ZnO nanorods. Sci. Rep. 8, 13722. https://doi.org/10.1038/s41598-018-32127-5.

Sappia, L., Felice, B., Sanchez, M.A., Martí, M., Madrid, R., Pividori, M.I., 2019. Electrochemical sensor for alkaline phosphatase as biomarker for clinical and in vitro applications. Sensor. Actuator. B Chem. 281, 221–228.

Schlimpert, M., Reichardt, W., Baeuerle, N., Hess, M.E., Metzger, P., Boerries, M., Peters, C., Kammerer, B., Brummer, T., Steinberg, F., Cathepsin, T.R., 2020. Deficiency in mammary epithelium transiently stalls breast cancer by interference with mTORC1 signaling. Nat. Commun. 11, 5133. https://doi.org/10.1038/s41467-020-18935-2.

Soldatkina, O.V., Soldatkin, O.O., Kasap, B.O., Kucherenko, D.U., Kucherenko, I.S., Kurc, B.A., Dzyadevych, S.V., 2017. A novel amperometric glutamate biosensor based on glutamate oxidase adsorbed on silicalite. Nanoscale Res. Lett. 12 (1), 260.

Song, M.J., Yun, D.H., Hong, S.I., 2009. An electrochemical Biosensor array for rapid detection of alanine amino transferase and aspartate aminotransferase. Biosci. Biotech. Biochem. 73. https://doi.org/10.1271/bbb.60043.

Takimura, Y., Saito, K., Okuda, M., Kageyama, Y., Saeki, K., 2007. Alkaliphilic *Bacillus* sp. Strain KSM-LD1 contains a record number of subtilisin-like serine protease genes. Appl. Microbiol. Biotechnol. 76, 395–405.

Van Waes, L., Lieber, C.S., 1977. Glutamate dehydrogenase: a reliable marker of liver cell necrosis in the alcoholic. Br. Med. J. 2, 1508–1510. https://doi.org/10.1136/bmj.2.6101.1508.

CHAPTER

Conclusion

13

Enzymes are one of the most wonderful, multifaceted molecules that are attracting desirable attention and acknowledgment in the industrial sector paralleled with consumer satisfaction and increase in its demand further. The pioneering work was done by biochemists, which paved the pathway for enzymology as a distinct branch. Chapter 1 enlists the main contributions of noble laureates that have resulted in the present-day status and development regarding the enzymes. Eminent chemists also had their valuable contribution in developing a clear concept regarding the physiochemical characteristics and their critical features that led to the study of its mechanism in detail. This chapter describes the important feature of enzymes as molecules with special reference to the nature of enzyme as a molecule, specificity, flexibility, and the models explaining these features. The chapter will help the readers to understand enzymes, their nature, and the grounds of gaining industrial recognition worldwide as an eco-friendly bioconversion tool for the production of a wide variety of products and processes.

The chapter also creates a platform for the readers to look into the future prospects of enzymes based on their present development. The hindrances in future research are also discussed with regard to the molecular restrictions the enzyme nature posses. Although to make the enzymes industrially feasible is the need of the present market, yet unresolved issues like enzyme sensitivity and inactivation in harsh environmental conditions of industrial processes are to be addressed in upcoming research.

Chapter 2 explains the distribution and diversity in microbial enzymes. In detail, the fundamental of enzyme classification and its basis are discussed using reactions that help in its classification. The chapter has enriched the latest classification with the inclusion of Class 7 of enzyme classification as according to the prior classification the Enzyme commission primarily classified enzymes into six classes on the basis of their ability to catalyze biochemical reactions. The results of studies on exploring the evolution of enzyme and its recognized function are included. The approach towards enzymes is now totally changed in which the amino acid sequence and their genetic makeup that coded them in relation to 3-dimensional structure and functions are studied holistically.

Further, the chapter leads to the extremophilic enzymes, their nature, and diversity in relation to their occurrence and microbial source. Studies based on the diversity of enzymes, their function, and relation to its ecological occurrence are explored in this chapter elucidating the cellular adaptations in extremophilic

Protocols and Applications in Enzymology. https://doi.org/10.1016/B978-0-323-91268-6.00012-0

organism. The outstanding features of extremophilic enzymes are discussed, which confer them with qualities like natural resistance toward physical and chemical environmental extremes prevailing during the environmental process.

In Chapter 3, the screening of potential microbes for enzymes of industrial significance is extensively discussed. The initial part of the chapter discloses history and dependency of humans on microbes for the production of quality products in food, beverages, and textiles. The chapter describes the various sources of enzymes and techniques of their isolation. The methods of isolating the microbes, the sources of microbes, various techniques of enrichment, and isolation of industrially important organisms are discussed. Industrially, extracellular enzymes are important and their screening techniques are developed for faster isolation of potential microbes. Enrichment culture technique, use of metagenomics, directed enzyme evolution, computational biology, throughput methods for screening of enzyme variants, agar plate method, and microtiter plate screening methods are discussed with their advantages and disadvantages.

High throughput methods are also discussed along with their limitations like fluorescence-activated cell sorting, in vitro compartmentalization, droplets, plasmid display, phage display, ribosome display and m-RNA display, the c-DNA display, reporter-based screening, and digital imaging.

Solid-state fermentation and submerged fermentation (SMF) for enzyme production are the two very basic approaches discussed in Chapter 4. Since human inception and civilization, the SMFs have accompanied humans. The benefits and disadvantages of SMF led to the search of an easier and economic method that leads to the development of solid-state fermentation. Although, solid-state fermentation has been used in traditional food preparations like cheese for a long time. The present scenario is explained in detail with the main components of the fermentation. The raw material, the microbe, and the bioreactor are thoroughly discussed with their interplay and final impact on product yield.

The various designs of bioreactors used in SMF are discussed with their limitations. The solid-state fermentation and the commonly used bioreactors are discussed in detail with the design and impact of the same on productivity and ease of the process.

With the background knowledge of the enzymes, their microbial sources, and the condition of enzyme production, Chapter 5 elaborates the detailed protocol for the assay of important industrial enzymes. The baseline of an enzyme assay procedure for any enzyme is to incubate the enzyme source with its substrate, and after a fixed incubation period, the reaction is terminated. The amount of substrate utilized or product formed is estimated by a suitable instrument. Protocols for assay of enzymes such as xylanases, fructosyltransferases, lipases, and proteases with all information for its performance in the laboratory are elaborated.

The assay of the enzyme helps to evaluate the quantity of enzyme present in the source and also the kinetics of the reaction. Chapter 6 helps to build the concept of the influence of various factors on enzyme activity that can be determined by the assay procedure learned in Chapter 5. The enzyme kinetics and its basic mathematical equations are elaborated. The energy transitions involved in enzyme-catalyzed reactions play an important role in the mechanism by which the enzymes catalyze the biotransformation at normal temperature. The most critical factors are Km, Vmax, and MM equation, and the factors affecting enzyme-catalyzed reactions are discussed with reference to Xylanases, Fructosyltransferases, lipases, and proteases that are desirous for industrial processes.

The understanding of kinetics and Km helps the researcher to scale up an enzyme and produce it at laboratory scale and industrial scale by selection of optimized parameters. In Chapter 7, the protocols of important industrial enzymes viz. small-scale and large-scale production of enzymes such as xylanases, fructosyltransferases, lipases, and proteases are elaborated.

The industries are looking forward to innovative strategies to improve enzyme production for industrial processes. Presently, enzyme engineering and genetic modifications are the key approaches to improvise the enzyme potential and increase the product yield to satisfy the economical expectations of industries. The current approaches for the selected enzymes are detailed in Chapter 8.

The past century has witnessed the evolution of enzymes from natural biomolecules to robust, tailor-made, industrially oriented biotransforming agents. With these drastic changes in the molecular capacity of natural enzymes, the scope and its relevant applications have also undergone an absolute revolution. The industrial processes and their demand are quite different from the actual optimum conditions in which enzyme performance is optimum. Nowadays, enzymes are becoming indispensable molecules due to the advantages they confer to the catalyzing process. The industrial applications in the present scenario and the relevance of enzymes in industries and the encountered problems are detailed in Chapter 9.

With the established advantages of enzymes as supreme catalytic molecules in an industrial scenario, the prospects of its productions are forwarding in directions that will make the enzyme production process economic. The traditional raw materials offer higher production costs as well as expensive downstream processing protocols. The concept of waste to value is now implemented for cheaper enzyme production. A thorough search of an appropriate combination of potential organisms and a suitable agroindustrial waste can be the best approach for economic enzyme production. The energy trapped in the wastes will be diverted to an economically important product rather than creating environmental chaos. The various wastes recycled for enzyme production are explained in Chapter 10 with reference to some industrially important enzymes.

With most of the applications discussed in other chapters, Chapter 11 mainly focuses on the significance of enzymes in the agricultural sector. Enzymes have an adorable performance in the agriculture sector also. The important enzymes successfully used for increasing fertility and productivity and as biopesticides and biomicrobiocides are discussed.

A protocol for laboratory-scale production of the enzymes as biofertilizers is appended so that field trials will help in better understanding of enzyme production, separation, and application.

Chapter 12 deals with biotechnological applications of enzymes prospective. Being an indispensable part of our daily lives, enzymes are still progressing in the industrial sector. Enzymes are now one of the best diagnostic tools used in the medical field. Enzymes are used in kits and assay protocols for disease detection. Enzymes are now popularly used as direct therapeutic agents for different disorders and deficiency diseases. The major demand in the industrial sector is of enzymes like proteases and lipases that find application in laundry and cleaning industries. All the enzyme-based products acknowledged by customers transmit the success story of the performance of this sensitive yet wonderful molecule.

Index

'*Note*: Page numbers followed by "f" indicate figures and "t" indicate tables.'

R

S